SpringerBriefs in Molecular Science

Green Chemistry for Sustainability

Series editor

Sanjay K. Sharma, Jaipur, India

Guangshan Zhu · Hao Ren

Porous Organic Frameworks

Design, Synthesis and Their Advanced
Applications

 Springer

Guangshan Zhu
State Key Laboratory of Inorganic
    Synthesis and Preparative Chemistry
Jilin University
Changchun
People's Republic of China

Hao Ren
State Key Laboratory of Inorganic
    Synthesis and Preparative Chemistry
Jilin University
Changchun
People's Republic of China

ISSN 2212-9898
SpringerBriefs in Molecular Science
ISBN 978-3-662-45455-8   ISBN 978-3-662-45456-5   (eBook)
DOI 10.1007/978-3-662-45456-5

Library of Congress Control Number: 2014955319

Springer Heidelberg New York Dordrecht London

Printed on acid-free paper

Springer-Verlag GmbH Berlin Heidelberg is part of Springer Science+Business Media
(www.springer.com)

Preface

Porous materials are of intensive academic and technological interest because of their vital applications for adsorbent, catalyst, ion exchanger, nanotechnology, etc. The development of porous materials has accompanied the demands of modern society. A large number of porous materials have been designed and synthesized in the past half century, including zeolite, mesoporous materials, metal-organic frameworks (MOFs) also known as coordination polymers, and porous organic frameworks (POFs). Although the above-mentioned porous materials seem to have major differences, they also possess corresponding consistencies. For zeolite, the basic structural unit is TO_4 tetrahedron. These primary building units (TO_4) are linked by corner sharing oxygen atoms together to form secondary building units (SBUs). SBUs can be connected in the form of cages or channels and finally lead to various zeolites with different structures. MOFs or coordination polymers are assembled by inorganic clusters and organic linkers. POF are constructed by purely organic units via robust covalent bonds. The synthesis procedure could be described as the assembly of building units via specific acting force.

The book *Porous Organic Frameworks: Design, Synthesis and Their Advanced Applications* is aimed at offering researchers with the most pertinent and up-to-date advances of POFs. The development of POF materials has attracted extensive attention thanks to their fascinating characteristics, such as structural designing-ability, high surface area, diverse pore dimensions, chemical functionalities, high chemical and thermal stabilities, etc. Encouragingly, POFs display excellent performances in the fields of gas storage, catalysis, host-guest chemistry, and optical and electronic properties, etc. The main benefit of this book is that it highlights the synthetic principles, and structural merits of most of the advanced POFs. In this book the important relationship between structures and functions of POFs is discussed. It is intended for scientists and researchers focusing on this research field. The material in this book will also benefit engineers concerning the applications of POFs. We would like to take this opportunity to thank Dr. Yuyang Tang, Qianjun Tang and Changhong Feng for their careful work in fixing grammar errors throughout this book. Special thanks go to Ye Yuan and Dr. Heping Ma, who contributed to the partial writing of Chaps. 3–6, including paperwork, figure design, etc.

August 2014   Guangshan Zhu
                                                                        Hao Ren

Contents

Chapter 1
Introduction to Porous Materials

Porous materials have been of intense scientific and technological interest because of their vital applications in adsorption, catalysis, ion exchange, nanotechnology, etc. [1]. Although people might not be familiar with the concept of "porous materials", they are widely used in our daily lives, such as in removal of harmful gases, dehumidizers, etc. What are porous materials? At microscopic scale, if the atoms of a material are not densely stacked but form voids, the material is defined as porous material. Nature is a magical world, which provides visual examples that help us to understand porous materials. For example, a honeycomb has hexagonal cells, seeming like a "house", and honeybees can "live" and get in and out of this building. The voids are formed by surrounding walls. The honeycomb could be regarded as the "host" and the honeybee the "guest" (Fig. 1.1).

Doubtless, the void of a honeycomb is very obvious with dimensions in centimeters and can be seen with our naked eyes. However, the voids of studied porous materials are at nanometer scales. According to the nomenclature recommended by the International Union of Pure and Applied Chemistry (IUPAC) [2], porous materials are classified as microporous materials (with pore diameters of less than 2 nm), mesoporous materials (with pore diameters of between 2 and 50 nm), and macroporous materials (with pore diameters of greater than 50 nm) based on their pore sizes. Besides, porous materials include inorganic materials (natural zeolites, synthetic zeolites, microporous aluminum phosphates, silica mesoporous materials, etc.), carbon-based materials, inorganic–organic hybrid materials metal–organic frameworks (MOFs), organic polymers, which are distinguished based on their structural and chemical compositions (Fig. 1.2).

When describing porous materials, several important structural characteristics should be clearly illustrated, including pore geometry, pore opening size, pore surface functionality, and polymeric framework structural features including composition, topology, and functionality (Fig. 1.3) [3]. Porosity is a profound parameter that describes porous materials. Nitrogen and Argon adsorption–desorption isotherms are the most important techniques to investigate the porous properties including the apparent surface area and pore size. The apparent surface area could

© The Author(s) 2015
G. Zhu and H. Ren, *Porous Organic Frameworks*, SpringerBriefs in Green Chemistry for Sustainability, DOI 10.1007/978-3-662-45456-5_1

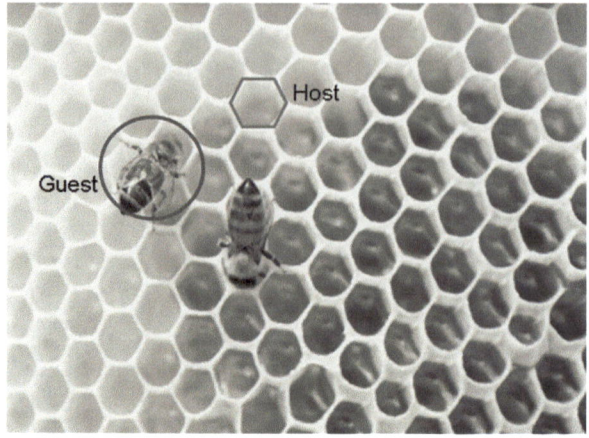

Fig. 1.1 Illustration of porosity existing in nature. The visual example is honeycomb

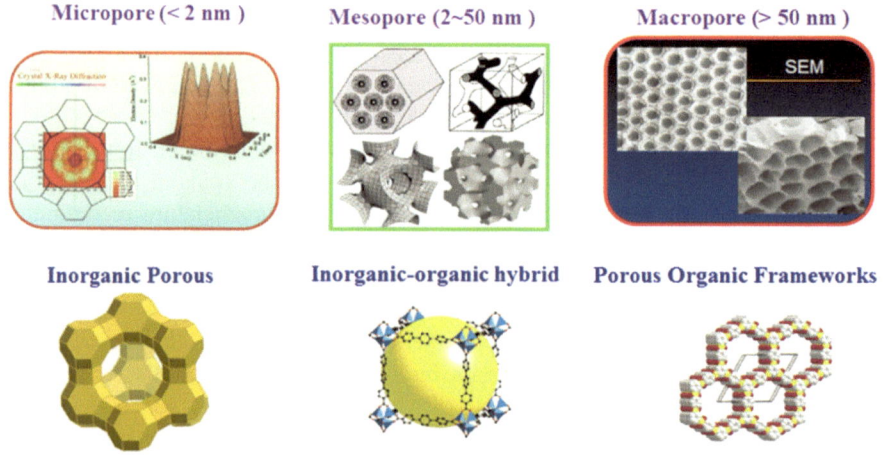

Fig. 1.2 Porous materials could be classified according to their pore size and their components

be calculated from Brunauer–Emmett–Teller (BET) model and Langmuir model, respectively. Commonly, the pore size distribution of porous materials is calculated by density functional theory (DFT) and nonlocal density functional theory (NLDFT). It should be mentioned that the model of adsorbent and the type of pores should be carefully selected to ensure the correctness and accuracy of analysis.

The development of porous materials is accompanied with the demands of society. A large number of porous materials have been designed and synthesized

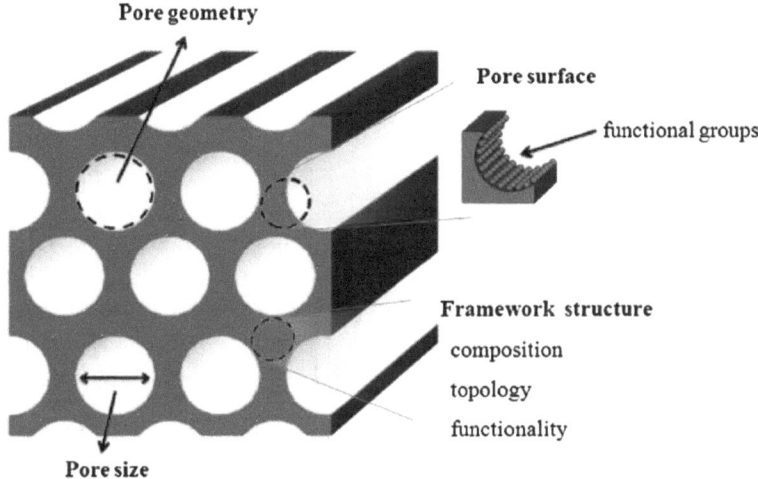

Fig. 1.3 Illustration of pore geometry, pore surface, pore size, and framework structure of porous polymers. Reprinted with permission from Ref. [3]. Copyright 2009, American Chemical Society

in the past half century. The investigation of porous materials experiences the following representative stages: (1) natural zeolites coined by Swedish mineralogist A.F. Cronstedt in 1756; (2) synthesis of zeolites under hydrothermal or solvothermal conditions; (3) report of the MCM family ordered mesoporous materials (MCM-41, MCM-48, etc.) in 1992 [4, 5]; (4) MOFs or coordination polymers through assembling inorganic units and organic linkers started at the beginning of the 1990s [6, 7]; (5) porous organic frameworks (POFs) constructed with purely light elements via robust covalent bonds [3, 8–18]. Although the above-mentioned porous materials seem to have major differences, they also possess certain similarities.

For zeolites, their basic structural unit is TO_4 tetrahedron, where TO_4 is $[SiO_4]^{4-}$ or $[AlO_4]^{5-}$. These primary building units (TO_4) are linked together by corner sharing oxygen atoms to form secondary building units (SBUs). SBUs can be connected in the form of cages or channels within the structure. Finally, constructing cages and rings of different sizes leads to various zeolites with different framework structures (Fig. 1.4).

Despite zeolites and microporous solids being widely applied in gas adsorption and separation catalysis, their pores are restrained to the subnanometer scale (<2.0 nm), limiting their application for larger molecules, especially for biological molecules in the areas of adsorption and catalysis. The report of MCM family materials with ordered mesopores has attracted considerable attention in the research field of mesoporous materials. In the process of synthesizing mesoporous materials, templates formed by the assembly of special molecules plays a vital

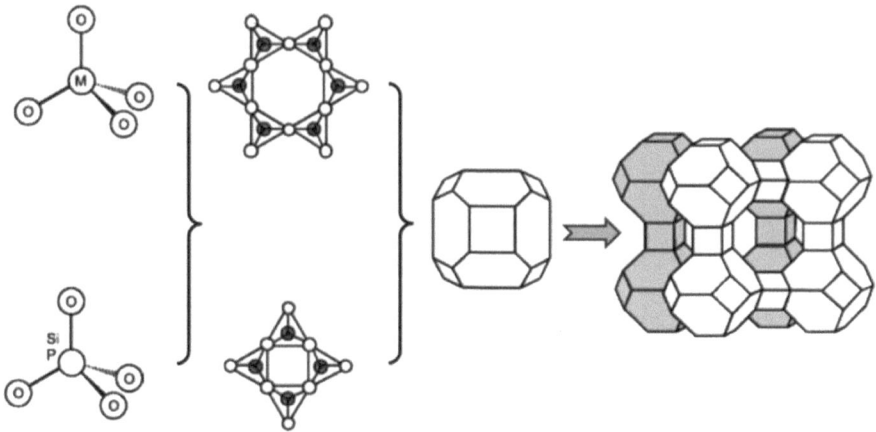

Fig. 1.4 Illustration of the formation of zeolite, from the primary TO_4 to secondary building units (SBUs), and SBUs further assemble to form extended zeolite

role, which can affect the final structure of mesoporous materials. The control of surfactant template is a fruitful strategy to design the pore and topology of mesoporous materials. Typical structures of mesoporous materials include [19]: (1) MCM-41, which has a two-dimensional hexagonal structure; (2) MCM-48, which has a cubic channel structure; (3) MCM-50, which has a lamellar structure.

Along with the development of purely inorganic porous materials, remarkable innovation for synthesis of porous materials has been made with the introduction of organic molecules as building composition of the structure, forming inorganic–organic hybrid compounds. Among these coordination compounds, MOFs, or coordination polymers constructed from metal ions or clusters as connectors and bridging organic ligands (Fig. 1.5) have been intensively investigated [6, 7]. In 1999, two robust MOFs, MOF-5 ($Zn_4O(bdc)_3$, bdc = 1,4-benzenedicarboxylate) [20] and HKUST-1 ($Cu_3(btc)_2$, btc = 1,3,5-benzenetricarboxylate) [21] considered as milestones in the development of MOFs, greatly promoted this fruitful research field. Compared with inorganic porous solids, MOFs have their intrinsic characters such as: (a) mild synthetic conditions, even at room temperature; (b) various organic ligands, which could be readily designed and modified; (c) abundant inorganic building connectors, metal ions, or metal clusters; (d) the structures and properties of MOFs could be tuned through pre-synthetic and post-synthetic design of organic ligands, inorganic SBUs, and synthetic conditions, etc.

In 2005, Yaghi et al. reported two 2D covalent–organic frameworks (COF) materials, COF-1 and COF-5 [23]. Their crystalline and nanoporous textures have changed researchers' understanding of porous materials and polymers. In fact, more than 40 years ago, hyper-cross-linked polymer (HCP) networks had already been successfully synthesized possessing amorphous and organic porous textures [24]. In this book, we introduce the concept of POFs [3, 8–18] constructed by purely light elements via robust covalent bonds, and present this new family

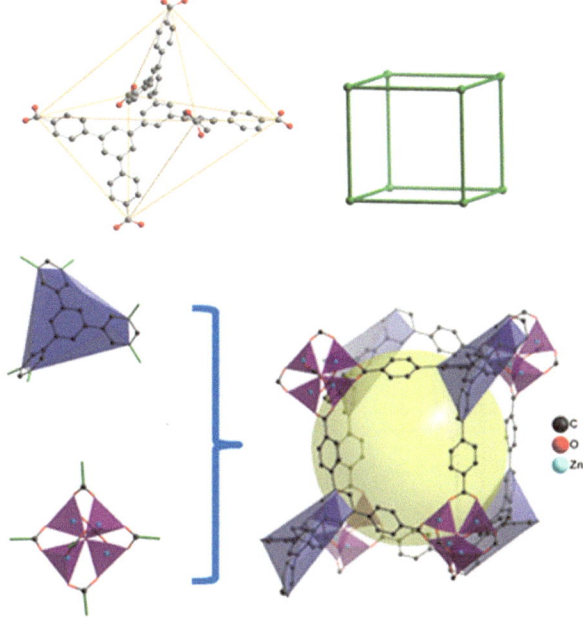

Fig. 1.5 Illustration of the formation of a MOF material, JUC-100. JUC-100 is constructed by octahedral organic ligand and metal–oxygen cluster with pcu topology. Reproduced from Ref. [22] with permission from The Royal Society of Chemistry

of porous materials. According to the criteria of structural regularity of porous materials, POF materials are sorted into two subclasses: amorphous porous materials and crystalline porous materials. POFs did not receive extensive attention before the emergence of COF-1 and COF-5. After that the development of POFs has greatly increased owing to their fascinating properties. Currently, POFs are receiving considerable attention thanks to their potential to merge the properties of both porous materials and polymers. First of all, most POFs display high stability even under rigorous conditions owing to the existence of strong covalent bonds. Secondly, they are generated by light elements (H, B, C, N, O, etc.) that may bring about high-specific surface area. Moreover, the structures and properties of POFs could be tuned by means of rational design and synthesis of building blocks. These unique characteristics of POFs guarantee that they have distinctive advantages. In contrast to inorganic materials, POFs exhibit higher surface area and are considered to be more designable and feasible to be modified. Compared with MOFs, POFs possess higher stability under rigorous conditions, such as high temperature, moisture, acid–base treatment, oxidative conditions, etc.

Investigations of POFs have been emerging as a new family of porous materials. When searching for the topic of "covalent–organic materials" on ISI Web of Knowledge, over 1,600 articles concerned with this subject were found until August 2014 (Fig. 1.6). It is obvious that the amount of articles has been greatly increasing in the past few years, indicating that it is an emerging and hot topic. In addition, the citation of articles also provides important information to evaluate the degree of concern in the research field. In 2005, Yaghi et al. reported the construction of crystalline POFs, COF-1 and COF-5. This is a milestone in the development

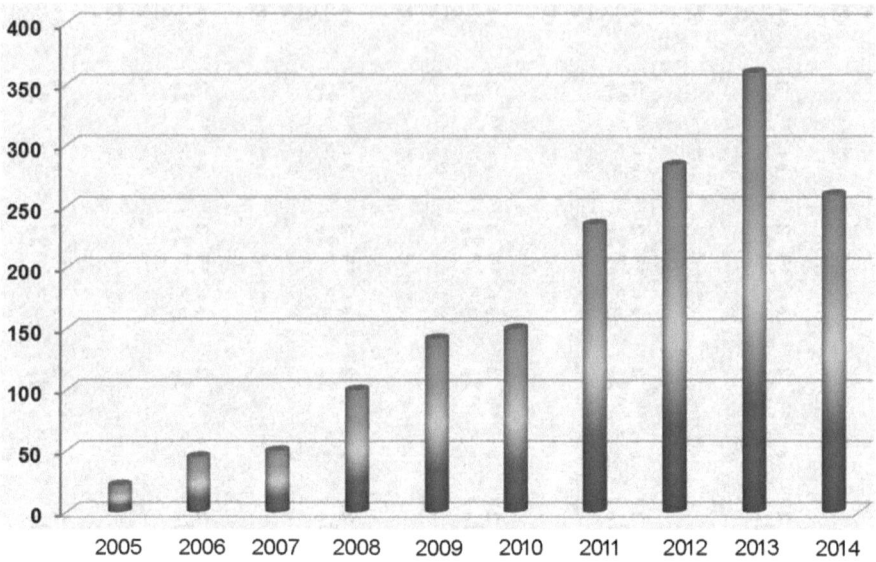

Fig. 1.6 The number of articles published between 2005 and 2014 regarding the topic of "covalent–organic materials", showing the increasing research interest. Data from ISI Web of Knowledge, Thomson Reuters, obtained by these keywords: "covalent–organic framework" OR "conjugated microporous polymer" OR "polymers intrinsic microporosity" OR "hyper-crosslinked polymers" or "porous aromatic framework" NOT "metal–organic framework". The data was collected by August 18th, 2014

of POF materials. In 2009, our group reported PAF-1 with high stability and exceptionally high surface area [25]. This work was highlighted by Cooper in Angew. Chem. Int. Ed [26]. In a reviewed article written by Cooper concerning ultrahigh surface area in porous solid, he stated that "PAF-1 is an important milestone in porous area since it demonstrates that long-range order is not necessarily a prerequisite for generating ultrahigh surface areas" [27]. Looking at the citation data of the COF-1, COF-5, and PAF-1 studies (Figs. 1.7 and 1.8), it is obvious that the investigation of POFs is growing rapidly in recent few years, further indicating that this is a hot topic in the research area of porous materials.

In the process of developing POF materials, there are many research groups contributing their enthusiasm. To be specific, we list here some prominent researchers from all over the world, including polymers intrinsic microporosity (PIMs) reported by McKeown and Budd [28–31], COFs reported by Yaghi [23, 32–34] (and other groups, such as Jiang [35–37], Bein [38], Dichtel [39–41], etc. [42–44]), conjugated microporous polymers (CMPs) reported by Cooper [45, 46], triazine-based organic frameworks (CTFs) reported by Thomas and Antonietti [47–49], covalent–organic polymers (COPs) reported by Cao [50, 51], HCPs reported by Tan [52, 53], and porous aromatic frameworks (PAFs) reported by Qiu and Zhu [25, 54–59], etc. [60–69]. With rapid development in the research of POF materials, diverse forms of POFs are being prepared.

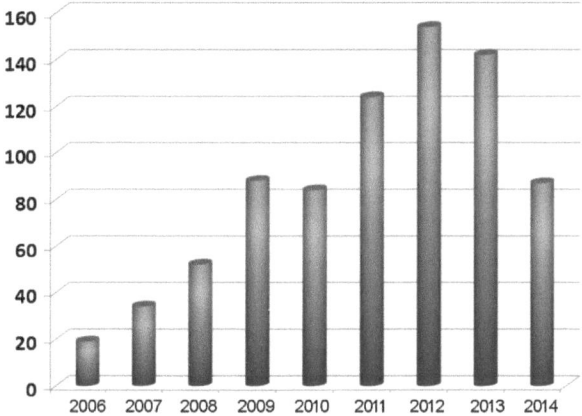

Fig. 1.7 The number of articles published between 2006 and 2014, citing the study reported by Yaghi et al. The first crystalline POFs, COF-1 and COF-5, were synthesized in 2005

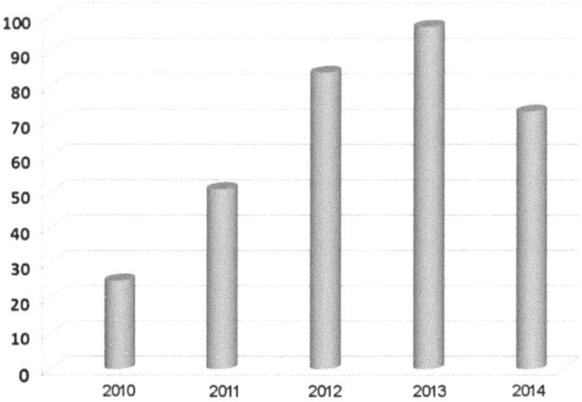

Fig. 1.8 The number of articles published between 2010 and 2014, which cited the study reported by Qiu and Zhu et al. PAF-1 was synthesized in 2009, which shows highest surface area among porous materials

As a promising family of porous materials, POF materials show fascinating characteristics such as structural design-abilities, high surface area, diverse porous dimensions, chemical functionalities, high chemical and thermal stabilities, etc. Therefore, these POF materials possessing high stability and high surface area are actively pursued as prospective materials for storage media, especially in the fields of clean energy and environmental control, such as H_2 and CH_4 storage and CO_2 capture. Moreover, abundant organic building blocks ensure that the structures and properties of POFs could be tuned through pre-synthesis design of constructing units, leading to special applications of POFs in different fields such as catalysis, host-guest chemistry, optics and electronics, etc. (Fig. 1.9).

Fig. 1.9 Advanced applications of POF materials, such as **a** gas storage, **b** catalyst, **c** host-guest chemistry, **d** optical and electronic materials

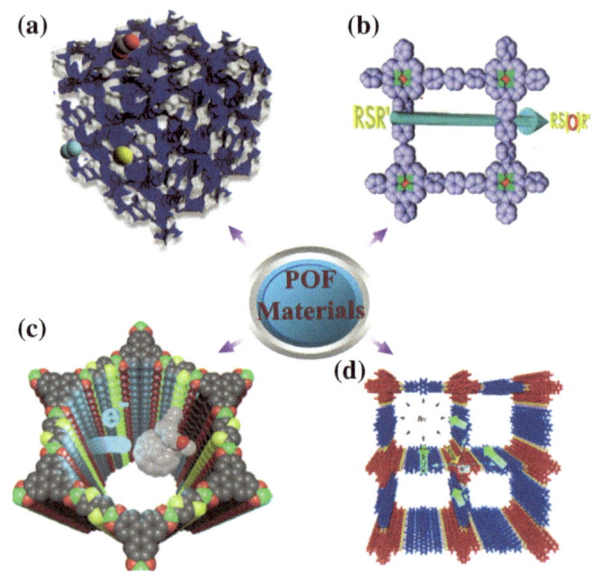

References

1. Xu R, Pang W et al (2004) Zeolite and porous materials chemistry
2. Rouquerol J, Avnir D, Faibrigde W et al (1994) Recommendations for the characterization of porous solids. Pure Appl Chem 66:1739–1758
3. Wu D, Xu F, Sun B et al (2012) Design and preparation of porous polymers. Chem Rev 112:3959–4015
4. Kresge C, Leonowicz M, Roth W et al (1992) Ordered mesoporous molecular-sieve synthesized by a liquid-crystal template mechanism. Nature 359:710–712
5. Monnier A, Schüth F, Huo Q et al (1993) Coopperative formation of inoraganic-organic interfaces in the synthsis of silicate mesostructures. Science 261:1299–1303
6. Eddaoudi M, Moler D, Li H et al (2001) Modular chemistry: secondary building units as a basis for the design of highly porous and robust metal-organic carboxylate frameworks. Acc Chem Res 34:319–330
7. Kitagawa S, Kitaura R, Noro S (2004) Functional porous coordination polymers. Angew Chem Int Ed 43:2334–2375
8. McKeown N, Budd P (2010) Exploitation of intrinsic microporosity in polymer-based materials. Macromolecules 43:5163–5176 (Some reviews concerning of POFs)
9. Dawson R, Cooper A, Adams D (2012) Nanoporous organic polymer networks. Prog Polym Sci 37:530–563
10. Feng X, Ding X, Jiang D (2012) Covalent organic frameworks. Chem Soc Rev 41:6010–6022
11. Kaur P, Hupp J, Nguyen S (2011) Porous organic polymers in catalysis: opportunities and challenges. ACS Catal 1:819–835
12. Zou X, Ren H, Zhu G (2013) Topology-directed design of porous organic frameworks and their advanced applications. Chem Commun 49:3925–3936
13. Vilela F, Zhang K, Antonietti M (2012) Conjugated porous polymers for energy applications. Energy Environ Sci 5:7819–7832
14. Ding S, Wang W (2013) Covalent organic frameworks (COFs): from design to applications. Chem Soc Rev 42:548–567

15. Zhang Y, Riduan S (2012) Functional porous organic polymers for heterogeneous catalysis. Chem Soc Rev 41:2083–2094
16. Xu Y, Jin S, Xu H et al (2013) Conjugated microporous polymers: design, synthesis and application. Chem Soc Rev 42:8012–8031
17. Xiang Z, Cao D (2013) Porous covalent–organic materials: synthesis, clean energy application and design. J Mater Chem A 1:2691–2718
18. Kalidindi S, Fischer R (2013) Covalent organic frameworks and their metal nanoparticle composites: prospects for hydrogen storage. Phys Status Solidi 250:1119–1127
19. Selvam P, Bhatia S, Sonwane C (2001) Recent advances in processing and characterization of periodic mesoporous MCM-41 silicate molecular sieves. Ind Eng Chem Res 40:3237–3261
20. Li H, Eddaoudi M, O'Keeffe M, Yaghi O (1999) Design and synthesis of an exceptionally stable and highly porous metal-organic framework. Nature 402:276–279
21. Chui S, Lo S, Charmant J et al (1999) Chemically functionalizable nanoporous material $[Cu_3(TMA)_2(H_2O)_3]n$. Science 283:1148–1150
22. Jia J, Sun F, Fang Q et al (2011) A novel low density metal-organic framework with pcu topology by dendritic ligand. Chem Commun 47:9167–9169
23. Cote A, Benin A, Ockwig N et al (2005) Porous, crystalline, covalent organic frameworks. Science 310:1166–1170
24. Xu S, Luo Y, Tan B (2013) Recent development of hypercrosslinked microporous organic polymers. Macromol Rapid Commun 34:471–484
25. Ben T, Ren H, Ma S et al (2009) Targeted synthesis of a porous aromatic framework with high stability and exceptionally high surface area. Angew Chem Int Ed 48:9457–9460 (Angew Chem 121:9621–9624)
26. Trewin A, Cooper A (2010) Porous organic polymers: distinction from disorder? Angew Chem Int Ed 49:1533–1535
27. Holst J, Cooper A (2010) Ultrahigh surface area in porous solids. Adv Mater 22:5212–5216
28. Budd P, Ghanem B, Makhseed S et al (2004) Polymers of intrinsic microporosity (PIMs): robust, solution-processable, organic nanoporous materials. Chem Commun 230–231
29. Budd P, Elabas E, Ghanem B et al (2004) Solution-processed, organophilic membrane derived from a polymer of intrinsic microporsity. Adv Mater 16:456–460
30. McKeown N, Ghanem B, Msayib K et al (2006) Towards polymer-based hydrogen storage materials: engineering ultramicroporous cavities within polymers of intrinsic microporosity. Angew Chem Int Ed 45:1804–1807
31. Carta M, Malpass-Evans R, Croad M et al (2013) An efficient polymer molecular sieve for membrane gas separations. Science 339:303–307
32. El-Kaderi H, Hunt J, Mendoza-Cortes J et al (2007) Designed synthesis of 3D covalent organic frameworks. Science 316:268–272
33. Uribe-Romo F, Hunt J, Furukawa H et al (2009) A crystalline imine-linked 3-D porous covalent framework. J Am Chem Soc 131:4570–4571
34. Furukawa H, Yaghi O (2009) Storage of hydrogen, methane, and carbon dioxide in highly porous covalent organic frameworks for clean energy applications. J Am Chem Soc 131:8875–8883
35. Wan S, Guo J, Kim J et al (2009) A photoconductive covalent organic framework: self-condensed arene cubes composed of eclipsed 2D polypyrene sheets for photocurrent generation. Angew Chem Int Ed 48:5439–5442
36. Nagai A, Chen X, Feng X et al (2013) A squaraine-linked mesoporous covalent organic framework. Angew Chem Int Ed 52:3770–3774
37. Xu H, Jiang D (2014) Covalent organic frameworks crossing the channel. Nat Commun 6:564–566
38. Dogru M, Handloser M, Auras F et al (2013) A photoconductive thienothiophene-based covalent organic framework showing charge transfer towards included fullerene. Angew Chem Int Ed 52:2920–2924

39. Spitler E, Dichtel W (2010) Lewis acid-catalysed formation of two-dimensional phthalocyanine covalent organic frameworks. Nat Chem 2:672–677

40. Spitler E, Colson J, Uribe-Romo F et al (2012) Lattice expansion of highly oriented 2D phthalocyanine covalent organic framework films. Angew Chem Int Ed 51:2623–2627

41. Smith B, Dichtel W (2014) Mechanistic studies of two-dimensional covalent organic frameworks rapidly polymerized from initially homogenous conditions. J Am Chem Soc 136:8783–8789

42. Tilford R, Mugavero S, Pellechia P et al (2008) Tailoring microporosity in covalent organic frameworks. Adv Mater 20:2741–2746

43. Rabbani M, Sekizkardes A, Kahveci Z et al (2013) A 2D mesoporous imine-linked covalent organic framework for high pressure gas storage applications. Chem Eur J 19:3324–3328

44. Biswal B, Chandra S, Kandambeth S et al (2013) Mechanochemical synthesis of chemically stable soreticular covalent organic frameworks. J Am Chem Soc 135:5328–5331

45. Jiang J, Su F, Trewin A et al (2007) Conjugated microporous poly(aryleneethynylene) networks. Angew Chem Int Ed 46:8574–8578

46. Jiang J, Su F, Trewin A et al (2008) Synthetic control of pore dimension and surface area in conjugated microporous polymer and copolymer networks. J Am Chem Soc 130:7710–7720

47. Kuhn P, Antonietti M, Thomas A (2008) Porous covalent triazine-based frameworks prepared by ionothermal synthesis. Angew Chem Int Ed 47:3450–3453

48. Kuhn P, Forget A, Su D et al (2008) From microporous regular frameworks to mesoporous materials with ultrahigh surface area: dynamic reorganization of porous polymer networks. J Am Chem Soc 130:13333–13337

49. Kuhn P, Thomas A, Antonietti M (2009) Toward tailorable porous organic polymer networks: a high-temperature dynamic polymerization scheme based on aromatic nitriles. Macromolecules 42:319–326

50. Zhou Y, Xiang Z, Cao D et al (2013) Covalent organic polymer supported palladium catalysts for CO oxidation. Chem Commun 49:5633–5635

51. Xiang Z, Cao D, Huang L et al (2014) Nitrogen-doped holey graphitic carbon from 2D covalent organic polymers for oxygen reduction. Adv Mater 26:3315–3320

52. Li B, Guan Z, Wang W et al (2012) Highly dispersed Pd catalyst locked in knitting aryl network polymers for Suzuki-Miyaura coupling reactions of aryl chlorides in aqueous media. Adv Mater 24:3390–3395

53. Luo Y, Li B, Wang W et al (2012) Hypercrosslinked aromatic heterocyclic microporous polymers: a new class of highly selective CO_2 capturing materials. Adv Mater 24:5703–5707

54. Ren H, Ben T, Wang E et al (2010) Targeted synthesis of a 3D porous aromatic framework for selective sorption of benzene. Chem Commun 46:291–293

55. Ren H, Ben T, Sun F et al (2011) Synthesis of a porous aromatic framework for adsorbing organic pollutants application. J Mater Chem 21:10348–10353

56. Zhao H, Jin Z, Su H et al (2011) Targeted synthesis of a 2D ordered porous organic framework for drug release. Chem Commun 47:6389–6391

57. Ma H, Ren H, Zou X et al (2013) Novel lithium-loaded porous aromatic framework for efficient CO_2 and H_2 uptake. J Mater Chem A 1:752–758

58. Yuan Y, Sun F, Zhang F et al (2013) Targeted synthesis of porous aromatic frameworks and their composites for versatile, facile, efficacious, and durable antibacterial polymer coatings. Adv Mater 25:6619–6625

59. Yuan Y, Sun F, Li L et al (2014) Porous aromatic frameworks with anion-templated pore apertures serving as polymeric sieves. Nat Commun 5:4260–4267

60. Zhu Y, Long H, Zhang W (2013) Imine-linked porous polymer frameworks with high small gas (H_2, CO_2, CH_4, C_2H_2) uptake and CO_2/N_2 selectivity. Chem Mater 25:1630–1635

61. Rabbani M, El-Kaderi H (2012) Department of template-free synthesis of a highly porous benzimidazole-linked polymer for CO_2 capture and H_2 storage. Chem Mater 23:1650–1653

62. Katsoulidis A, Kanatzidis M (2011) Phloroglucinol based microporous polymeric organic frameworks with OH functional groups and high CO_2 capture capacity. Chem Mater 23:1818–1824

63. Liu X, Li H, Zhang Y et al (2013) Enhanced arbon dioxide uptake by metalloporphyrinbased icroporous covalent triazine framework. Polym Chem 4:445–448
64. Bhunia A, Vasylyeva V, Janiak C (2013) From a supramolecular tetranitrile to a porous covalent triazine-based framework with high gas uptake capacities. Chem Commun 49:3961–3963
65. Shen C, Bao Y, Wang Z (2013) Tetraphenyladamantane-based microporous polyimide for adsorption of carbon dioxide, hydrogen, organic and water vapors. Chem Commun 49:3321–3323
66. Woodward R, Stevens L, Dawson R (2014) Swellable, water- and acid-tolerant polymer sponges for chemoselective carbon dioxide capture. J Am Chem Soc 136:9028–9035
67. Hauser B, Farha O, Exley J et al (2013) Thermally enhancing the surface areas of yamamoto-derived porous organic polymers. Chem Mater 25:12–16
68. Yuan D, Lu W, Zhao D et al (2011) Highly stable porous polymer networks with exceptionally high gas-uptake capacities. Adv Mater 23:3723–3725
69. A S, Zhang Y, Li Z et al (2014) Highly efficient and reversible iodine capture using a metalloporphyrin-based conjugated. Chem Commun 50:8495–8498

Chapter 2
Principles for the Synthesis of Porous Organic Frameworks

Abstract To date, hundreds of porous organic frameworks (POFs) have been successfully designed and synthesized. With regard to POFs, the first important problem is the constructing strategies of POFs. In this chapter, we introduce some principles for the synthesis of POF materials. Based on the type of POFs (crystalline or amorphous), the discussion of POFs is classified into two parts. For the synthesis of POF materials, two most vital factors should be followed. One is the monomers that are the fundamental constructing units for the synthesis of POFs, and the other is the polymerization reactions. The structure and property of POFs are determined by the utilization of monomers and polymerization reactions. In other words, the synergistic effect of two factors could influence the chemical and physical characteristics of the resulting samples.

Keywords Configuration · Reactive groups · Polymeric reactions · Crystalline POFs · Amorphous POFs

2.1 Introduction

In this chapter, we discuss the principles for the synthesis of porous organic frameworks (POFs). In POF material synthesis, there are two most vital factors that attract our attention. One is the monomers, which are the fundamental constructing units for the synthesis of POFs, and the other is the polymerization reactions. The structures and properties of POFs are greatly affected by the utilization of monomers and polymerization reactions. With regard to monomers, their corresponding reactive groups and their spatial configurations play extremely important roles, which determine the choice of polymerization reaction and the final structure of POFs (Fig. 2.1). In addition, based on the types of employed reactions, monomers are reasonably designed and prepared to satisfy corresponding demands. Therefore, in considering how to prepare POFs with permanent pores,

© The Author(s) 2015

G. Zhu and H. Ren, *Porous Organic Frameworks*, SpringerBriefs in Green Chemistry for Sustainability, DOI 10.1007/978-3-662-45456-5_2

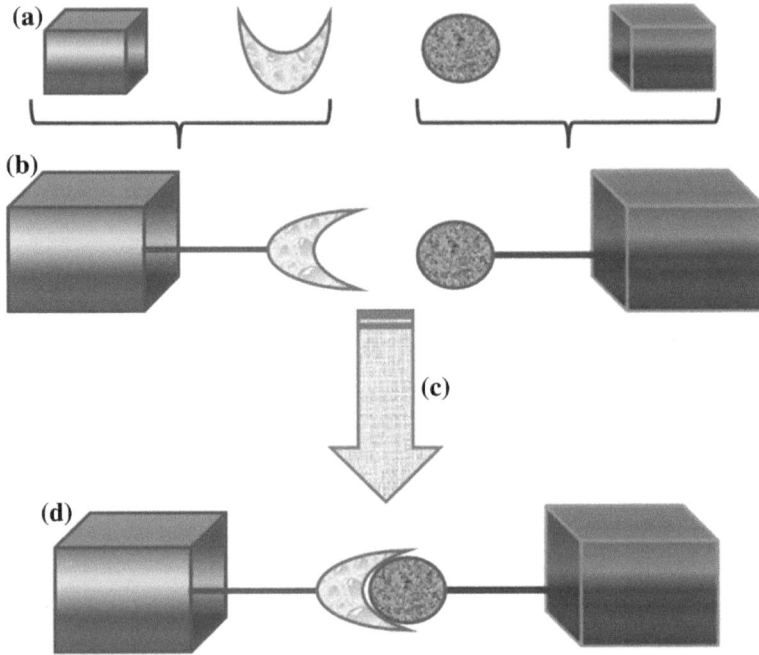

Fig. 2.1 Illustration of synthesis procedure of porous organic frameworks: **a** two parts of the monomers, which are the reactive groups and their configurations; **b** monomers with special reactive groups and configurations; **c** effective polymerization reactions based on the reactive groups; **d** porous organic frameworks constructed by the suitable building units via effective reactions

the monomers and reactions should be combined simultaneously. In other words, the synergistic effect of two factors could influence the chemical and physical characteristics of the resulting samples.

To present the resulted structures of POFs clearly and concretely, the underlying topology plays a fundamental role as schematically shown in Fig. 2.2 [1]. The basic and necessary components in POFs are the organic monomers, which are generally utilized in the construction of secondary building units (SBUs). Meanwhile, possible sites for forming bonds of the monomers could determine the connection manners, and they will direct the geometry of the dimensional nets, including one-, two-, and three-dimensional periodic nets. By the method of simplification/transformation, the monomers are transformed into corresponding nets, seen from the exemplified monomers in Fig. 2.2. Combined with SBUs and their connectivity, "true" topology of POFs is predicted to describe the most possible structure. Therefore, a comprehensive understanding of the underlying topologies will provide favorable assistance in the directionality for the construction of open and robust POF frameworks.

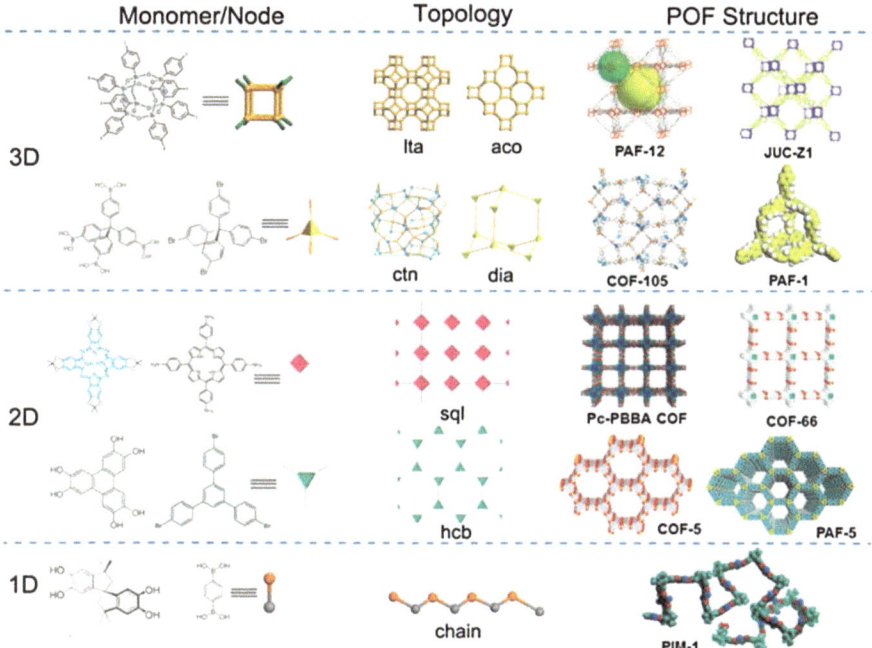

Fig. 2.2 Schematic illustration of building blocks, underlying topologies, and final structures of POF materials. Reproduced from Ref. [1] with permission from The Royal Society of Chemistry

2.2   General Strategy for the Synthesis of POFs

2.2.1   Bottom-Up or Top-Down Bidirectional Way for Construction of POFs

At the primary stage of investigations of POFs, their preparations are performed by serendipity. To avoid the trial-and-error way, a facile and rational strategy is greatly demanded for the design and synthesis of POF materials systematically. Based on this consideration, a topology-directed approach could realize the feasibility of synthesizing predetermined POF materials for specific purposes. Combining our research with other representative investigations, we have proposed bottom-up or top-down bidirectional way for construction of POFs (Fig. 2.3). In the bottom-up way, the organic monomers are originally fixed. By carefully analyzing the configuration of available nodes, the preferred topology would be determined first and the structures of POFs could be predicted subsequently with the assistance of computational methods and tools (e.g., Materials Studio software). Similarly, in the top-down method, the objective POF structure is specifically designed. By utilization of the computing

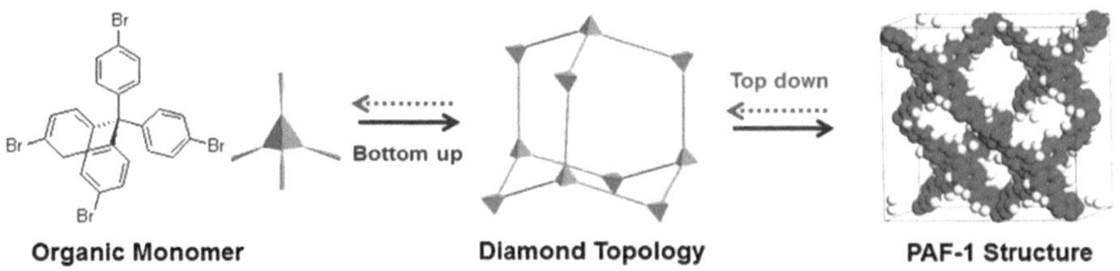

| **Organic Monomer** | **Diamond Topology** | **PAF-1 Structure** |

Fig. 2.3 Topology-directed synthesis of POF material via bottom-up and top-down approaches, taking PAF-1 as example. Reproduced from Ref. [1] with permission from The Royal Society of Chemistry

skills and RCSR database organized by Yaghi and O'Keeffe (http://rcsr.anu.edu. au/), the underlying topological possibilities could be produced. Subsequently, the suitable monomers would be selected and prepared to satisfy the demands. From this point of view, a topology-based strategy offers guidelines for the fabrication of POF architectures with interesting structural properties aiming at diverse applications.

2.2.2 The Configuration of Monomers

According to the geometry of POF materials, they could be classified into three types (one-, two-, and three-periodic nets). The final structures of POF materials are directly decided by the original configurations of the monomers. To construct POFs with high porosity, the monomers should have special configuration such as planar triangle, octahedron, linear pattern, tetrahedron, and square, etc. (Fig. 2.4).

The monomers listed in Fig. 2.5 could meet the requirements that are able to satisfy the demands on the configuration of building units. It has been documented that those monomers with the above-mentioned structures present good abilities for constructing POFs.

Through selecting suitable building units and effective strategies, we could obtain POF materials. The next step should be the determination of the structure and porosity of POFs. Therefore, POFs are characterized by Fourier transform infrared spectroscopy (FT-IR), ^{13}C solid-state NMR, powder X-ray diffraction (PXRD), scanning electron microscopy (SEM), transmission electron microscopy (TEM), thermogravimetric analysis (TGA), and N_2 gas sorption. According to the pattern of PXRD, POFs are distinguished into two types, ordered crystalline POFs or randomly amorphous POFs. To date, hundreds of POFs have been successfully designed and synthesized. In the following sections, POFs are discussed in accordance with whether they are crystalline or not.

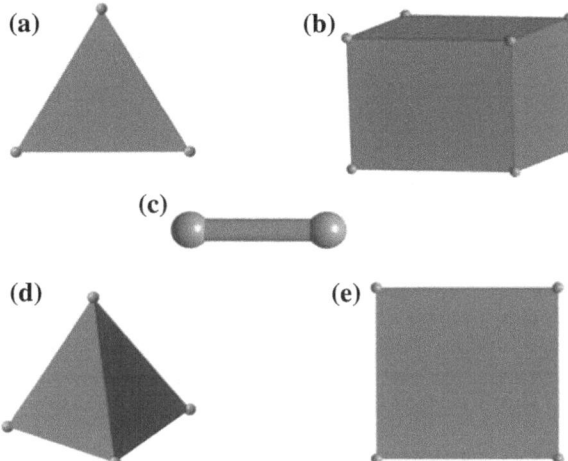

Fig. 2.4 The proposed configurations of the monomers include *planar triangle*, *octahedron*, *linear pattern*, *tetrahedron*, and *square*

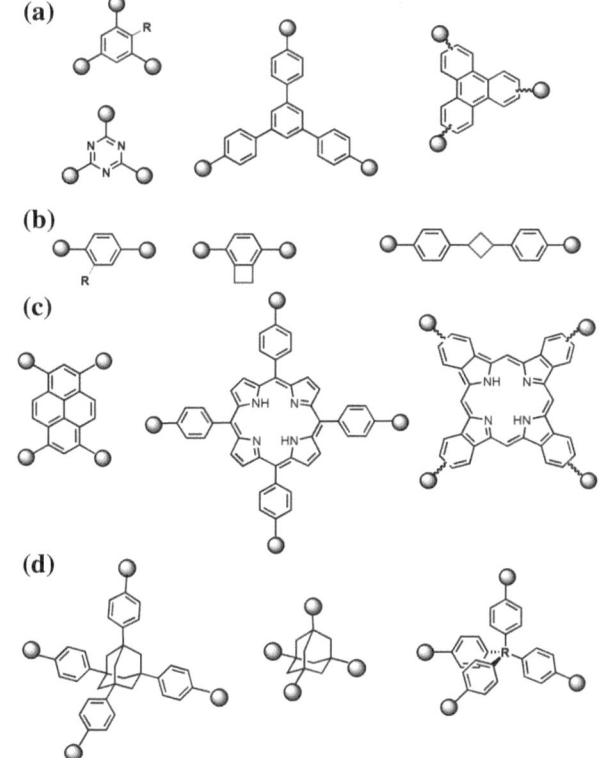

Fig. 2.5 Typical monomers with regular configurations, the reactive groups could be tuned according to the polymerization reactions

2.3 Crystalline POFs

As a family of POF materials, hyper-cross-linked polymers (HCPs) emerged more than 40 years ago. Before 2005, relatively few POFs were synthesized and reported. One type of POFs is the representative polymers intrinsic microporosity (PIMs) reported by McKeown and Budd [2–4]. At the same time, chemists and materials scientists believed POFs should be amorphous rather than crystalline. In 2005, an unexpected example of POFs reported by Yaghi and coworkers had thoroughly changed this traditional concept [5]. Using effective topological design principle, crystalline porous organic frameworks, referred as covalent organic frameworks (COFs), had been obtained by periodic connection via covalent bonds. COFs are the first successful examples of covalent crystalline POFs, which realize the ordered arrangement of sole organic building units with atomic precision.

2.3.1 Dynamic Covalent Chemistry

Nowadays, different types of COFs have been developed through the rational design of reactions and building blocks. Which reactions and building units are favorable to construct crystalline COF? Earlier, amorphous POFs were generally synthesized through various coupling reactions, which are kinetically controlled reactions and would lead to irreversible formation of covalent bonds. This irreversibly formed covalent bonding leads to the formation of randomly arranged structures which are hardly to be adjusted. Therefore, it is difficult to synthesize crystalline POFs via kinetically controlled reactions.

As the first example of crystalline POFs, why can COF-1 and COF-5 be successfully prepared? Analysis of the formation of COF-1 and COF-5 procedure is useful to explore the reasons. Figure 2.6 shows the condensation reactions for the synthesis of COF-1 and COF-5, which are based on the molecular dehydration reaction [5]. For COF-1, three boronic acid molecules converge to form a planar six-membered B_3O_3 (boroxine) ring with the elimination of three water molecules, and a honeycomb-like structure is expected to form using 1,4-benzenediboronic acid (BDBA) as monomers. For COF-5, an analogous condensation reaction is employed, which forms borate ester. First, the dehydration reaction between boronic acid and diol generates a five-membered borate ester ring (BO_2C_2). Then, it is found that the entire coplanar extends to a sheet structure.

Similarly, the reactions of borate anhydride formation and borate ester formation are dynamic covalent chemistry (DCC), resulting in the reversible formation of covalent bonds. In other words, the networks of COF-1 and COF-5 could be formed, broken, and reformed to finally obtain a stable state. Compared with kinetically

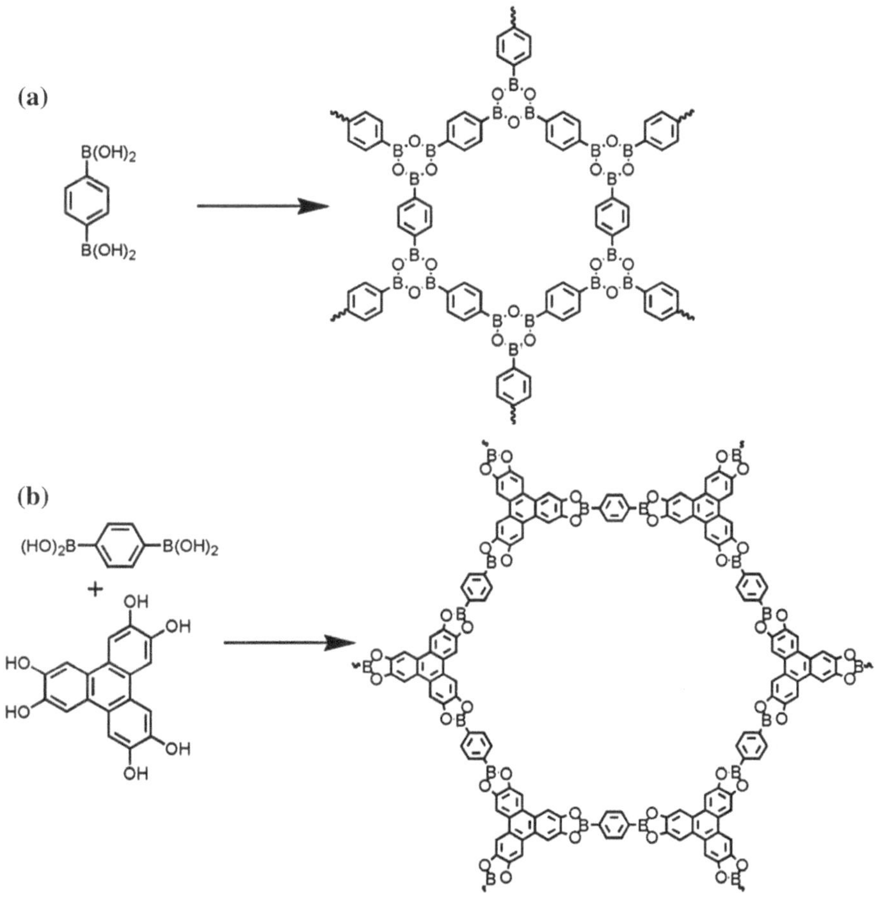

Fig. 2.6 The synthesis procedure of COF-1 and COF-5

controlled reactions, DCC is thermodynamically controlled. When forming COF materials, DCC reactions experience an "error checking" or "proof-reading" process. This process allows the formed structure to adjust itself to reduce its structural defects and form a stable state. Therefore, crystalline COFs would finally form with the most thermodynamically stable structures, which are conducted by the reversible reaction systems.

The COF-1 and COF-5 have provided strong evidence that DCC is favorable to construct crystalline POFs. By using DCC strategy, other dynamic chemical reactions have been proposed to construct COFs, which are indicated in Fig. 2.7 [6, 7]. According to the reactions, the reactive groups such as –B(OH)$_2$, diols, –CHO, –NH$_2$, –CN, etc., are preferred.

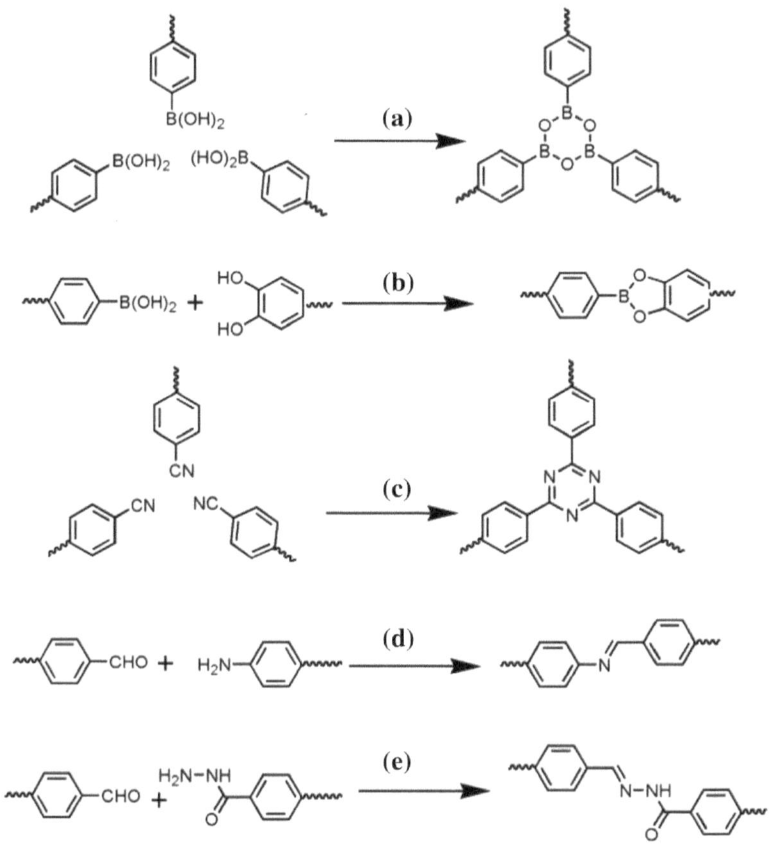

Fig. 2.7 Examples of the dynamic chemical reactions for the preparation of COFs. **a** Boronate ester ring formation. **b** BO_2C_2 ring formation. **c** $ZnCl_2$ mediated nitrile cyclotrimerization. **d** Imine reaction. **e** Hydrazone reaction

2.3.2 Topology Concerns in COFs

COFs are organic porous polymers with periodic networks. To clearly illustrate the structures of COFs, the concept of topology, which is a clear symbol describing framework structures directly, has been proposed. The understanding of topology plays a fundamental role in bridging the various building blocks and the possibly resulted structures. Topology is the mathematical study of shapes and topological spaces. It is a subfield of mathematics concerned with the properties of space that are preserved under continuous deformations, including properties such as connectivity, continuity, and boundary. In virtue of analysis of concepts such as space, dimension, and transformation, topology theory has been developed as a field of studying the geometry and set theory.

To be specific, a topology is given as a set of ordered pairs of the vertices on a net, which is applied to illustrate how elements of a set are related spatially to each other. Besides, each pair determines an edge of the net. Topology has been applied to modern chemistry, particularly in the field of crystallography. Description of metal-organic frameworks (MOFs) using different topologies has a special advantage, which simplifies the complex structures of MOFs. How does topology theory describe and organize the structure (atomic arrangement) of MOFs? Metal ions/clusters and organic ligands are two components of MOFs. When they are regarded as well-defined molecular clusters, MOFs could be assembled from these building units with simple geometrical shapes. Many of the most common nets are collected in a searchable database and identified by widely used RCSR symbols, which is carefully developed by Prof. Yaghi and O'Keeffe (http://rcsr.anu.edu.au/). The reticular chemistry with the same concept of topology-based science has been successfully extended to the synthesis of COFs.

According to the criteria defined by Prof. O'Keeffe, a series of two-dimensional and three-dimensional structures could be reasonably built based on the conformation of varied nodes. In order to achieve defined 2D topological structures, planar segments (linear, triangle, or square) are required. For example, a plane with hexagonal lattice could be formed by connecting the triangles; the equal squares are linked together to form a plane as the square lattice. When an extra linear linker exists, the lattice will be extended correspondingly. For clarity, the types of SBUs and corresponding structures, and their connectivity are displayed in Fig. 2.8. It is worth mentioning that if the monomers are linear units (Fig. 2.8a), in order to obtain extended networks, additional triangular nodes should be formed. Therefore, the boroxine ring formation and $ZnCl_2$ mediated nitrile cyclotrimerization could meet this requirement.

For construction of 3D frameworks, polyhedra are generally required. These polyhedra are connected with planar SBUs or linked to each other to form extended structures (Fig. 2.9). For example, Fig. 2.9 illustrates the constructing procedure of two 3D structures with the same building units. First, the elementary building units are tetrahedron and triangle. The planar triangular ring connects three tetrahedral units to form a secondary building unit. These building units are further connected together to produce the expanded ctn and bor structure, respectively.

Fig. 2.8 Planar constructing units and their corresponding structures

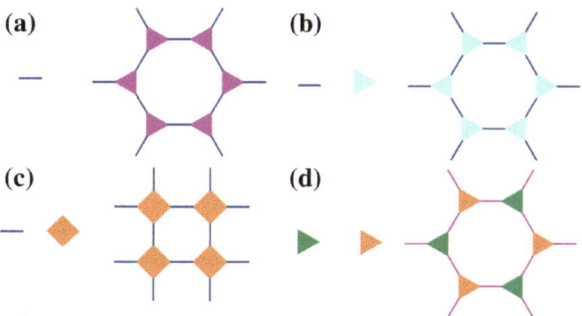

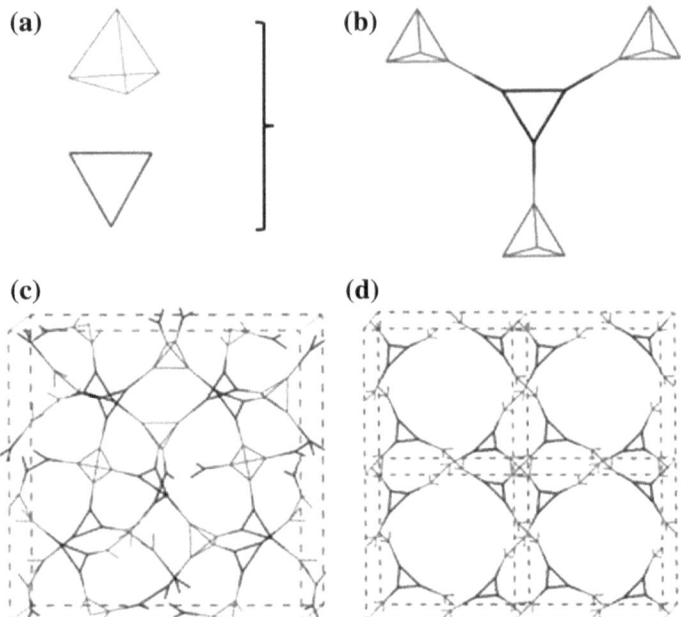

Fig. 2.9 A secondary building unit is formed by linking three tetrahedral units with the planar triangular ring, (**a**) and (**b**). These building units connected together can produce the expanded ctn (**c**) and bor (**d**) nets, respectively

2.3.3 Structure of Building Blocks

Besides the reasonable reactions, the structure of the building blocks should be elaborately designed to match the demand for production of crystalline and ordered COFs. The understanding of the DCC and the concept of topological structure will provide vital guidelines for the fabrication of COF architectures. According to this knowledge, the design and synthesis of COF materials could be more targeted. It is noted that the reactions and the monomers should be considered simultaneously. It has been documented that the reactions shown in Fig. 2.7 are effective to produce COFs, so these reactions are the best and primary choice. In addition, based on the reported reactions, the monomers should possess the following reactive groups, including $-B(OH)_2$, diols, $-CHO$, $-NH_2$, and $-CN$, which will trigger the dynamic covalent bond formation. Apart from the reactive groups, the structure of the building blocks must have suitable geometry, which could determine the final structure of COF. After careful consideration, the suitable candidates are selected to meet the requirement.

Compared with the relatively fixed functional groups, the geometries of the monomers are more complicated. Topology offers a powerful tool that could guide the synthesis of COFs. To obtain COFs with ordered and porous structures, generally speaking, the building blocks should be conformationally rigid, and the bond formation direction must be discrete. To satisfy these requirements, a series of building blocks are designed, which are distinguished according to their directional symmetry of the reactive groups.

How to select building units to construct COFs? Two requirements should be considered: one is that the monomers should contain the above-mentioned reactive groups which are suitable for DCC reactions; the other is that the geometries of monomers should match with the regular structures. Theoretically, the building blocks could be designed on the basis of the requirements. However, in the actual experimental procedure, the feasibility of preparing monomers should be considered. As simplified symmetry notation, these 2D or 3D monomers are referred to as 2D-C_2, 2D-C_3, 2D-C_4, and 3D-T_d.

Through carefully analyzing the geometry of monomers and the constructing reactions, the forming process of COFs could be discussed in detail. (1) Along with the reactions occur new types of connection forms; (2) the topological structures do not change during bond formation. Regarding (1), the reactions of borate anhydride formation and $ZnCl_2$ mediated nitrile cyclotrimerization could offer newly formed triangular connections. Thus, even the monomers are 2D-C_2 types, the resulting discrete linkers provide favorable conditions for the formation of COF networks. Therefore, for these 2D-C_2 types' relations, the reactive groups of the monomers are fixed on the borate acid and nitrile groups. As shown in Fig. 2.10, the representative samples are COF-1 and CTF-1 [8].

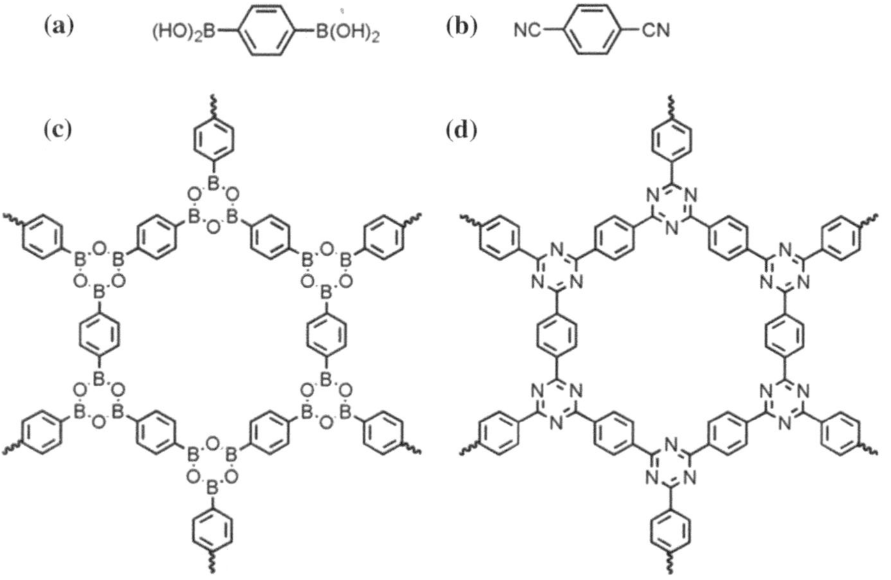

Fig. 2.10 Schematic representation of COF-1 and CTF-1

2.3.4 Structure of Crystalline POFs

Generally speaking, to afford 2D COFs with designed topology and pore structure, 2D blocks are the fundamental elements. Besides the above-mentioned reaction, most of the 2D COFs could be formed in terms of 2D-C_2 + 2D-C_3, 2D-C_3 + 2D-C_3, and 2D-C_2 + 2D-C_4. Figure 2.11 lists the representative building blocks with different geometries to design 2D COFs.

For porous materials, pore size is one of the most important features. The pore size of the resulted COF materials could be tuned through adjusting the length of monomers or decorating additional groups into the building units. Figures 2.12 and 2.13 indicate that the factors lead to the change in pore sizes [9, 10].

To obtain 3D COFs, one of the monomers must be 3D block, and the types of combinations are 3D-T_4 + 3D-T_4, 3D-T_4 + 2D-C_2, and 3D-T_4 + 2D-C_3. Figure 2.14 shows the employed 3D-T_4 monomers, such as tetraphenyl methane and tetraphenyl silicon derivatives, adamantane derivative, etc.

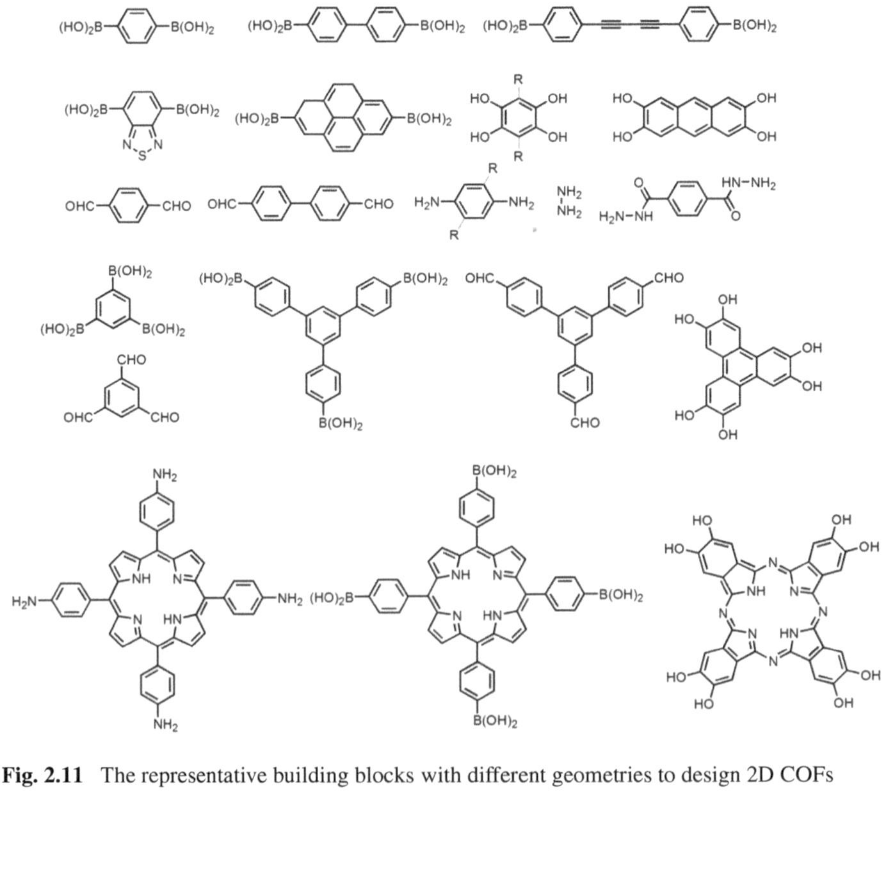

Fig. 2.11 The representative building blocks with different geometries to design 2D COFs

R = H, COF-18Å
R = CH_3, COF-16Å
R = CH_2CH_3, COF-14Å
R = $CH_2CH_2CH_3$, COF-11Å

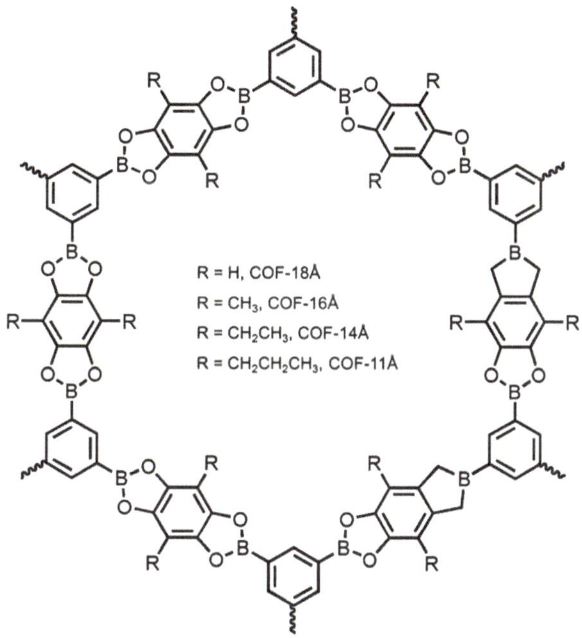

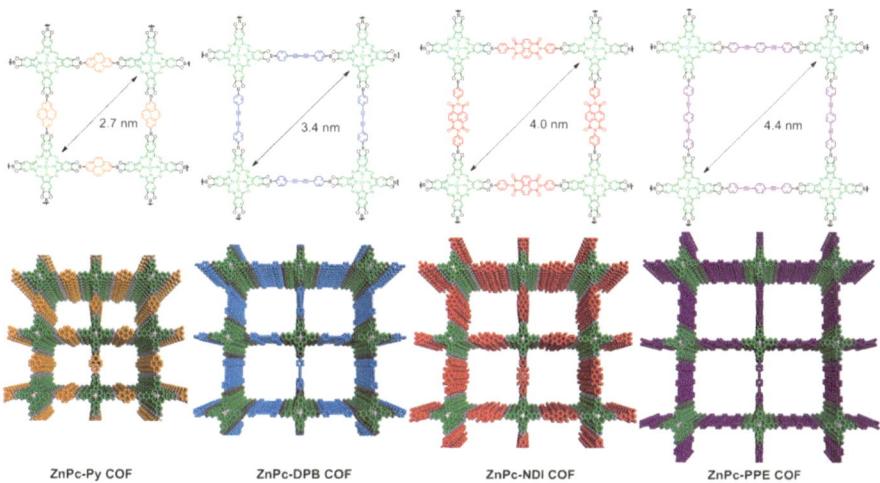

Fig. 2.12 Schematic representation of pore size of COFs tuned by the alkyl chains in the walls

2.7 nm
3.4 nm
4.0 nm
4.4 nm
ZnPc-Py COF
ZnPc-DPB COF
ZnPc-NDI COF
ZnPc-PPE COF

Fig. 2.13 Chemical and extended structures of the expanded ZnPc COFs. Each COF forms a two-dimensional layered network containing zinc phthalocyanines joined by (*left* to *right*) pyrene, diphenylbutadiyne, naphthalenediimide, and phenylbis(phenylethynyl) units. Reprinted with permission from Ref. [10]. Copyright 2012, Wiley-VCH

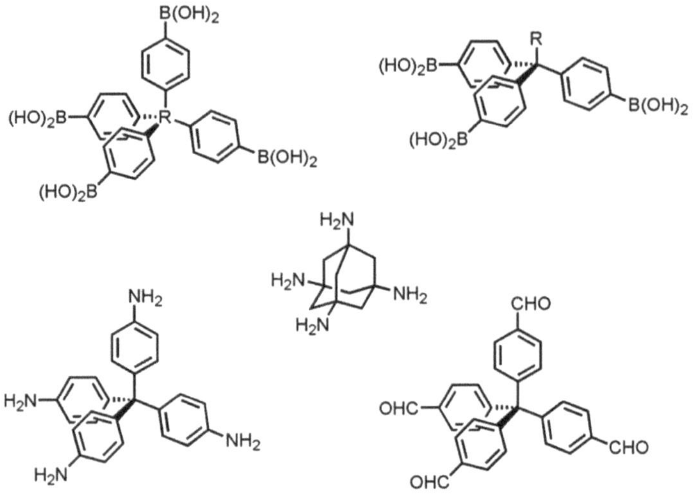

Fig. 2.14 The representative building blocks with different geometries to design 3D COFs

Compared with 2D COFs, 3D COFs are relatively rare owing to their limited building units. The following strategies are employed to construct 3D COFs: (1) the self-condensation of tetrahedral (3D-T_4) nodes; (2) co-condensation of tetrahedral nodes with linear (2D-C_2) or triangular (2D-C_3) building blocks. The pioneering work of rational design for 3D COFs was achieved by Yaghi et al. [11]. In the synthesis, tetra(4-dihydroxyborylphenyl)methane (TBPM), and tetra(4-dihydroxyborylphenyl) silane (TBPS) are selected as the tetrahedral nodes (Fig. 2.15). Dehydration reaction of borate acid produces triangular B_3O_3 ring. Via self-condensation of tetrahedral blocks and co-condensation with the triangular units, two preferred nets are envisioned, which are ctn and bor, respectively. The self-condensation of the tetrahedral TBPM and TBPS give COF-102 and COF-103, respectively, both of which are ctn topology. Additionally, hexahydroxytriphenylene (HHTP) could be used as triangular units. In contrast, the co-condensation of HHTP with TBPS forms COF-108 with a bor topology.

Another successful example of 3D COFs is illustrated by Yan et al. [12]. Combining a tetrahedral alkyl amine with planar triangular building units, 3D base-functionalized COFs, BF-COF-1, and BF-COF-2 are obtained (Fig. 2.16). PXRD analysis indicates that they are ctn topology. N_2 sorption shows that they are microprous. It is notable that they have excellent catalytic activities with high conversion yield owing to their highly efficient size selectivity.

Besides the ctn and bor type COFs, another developed type is dia topology. It is worth mentioning that COF-300 reported by Yaghi et al. is the first imine-linked 3D COF (Fig. 2.17) [13]. To form COF with dia topology, the basic building units are tetrahedral nodes. The condensation of the tetrahedral building block tetra-(4-anilyl) methane with the linear linking unit terephthaladehyde produces a material with

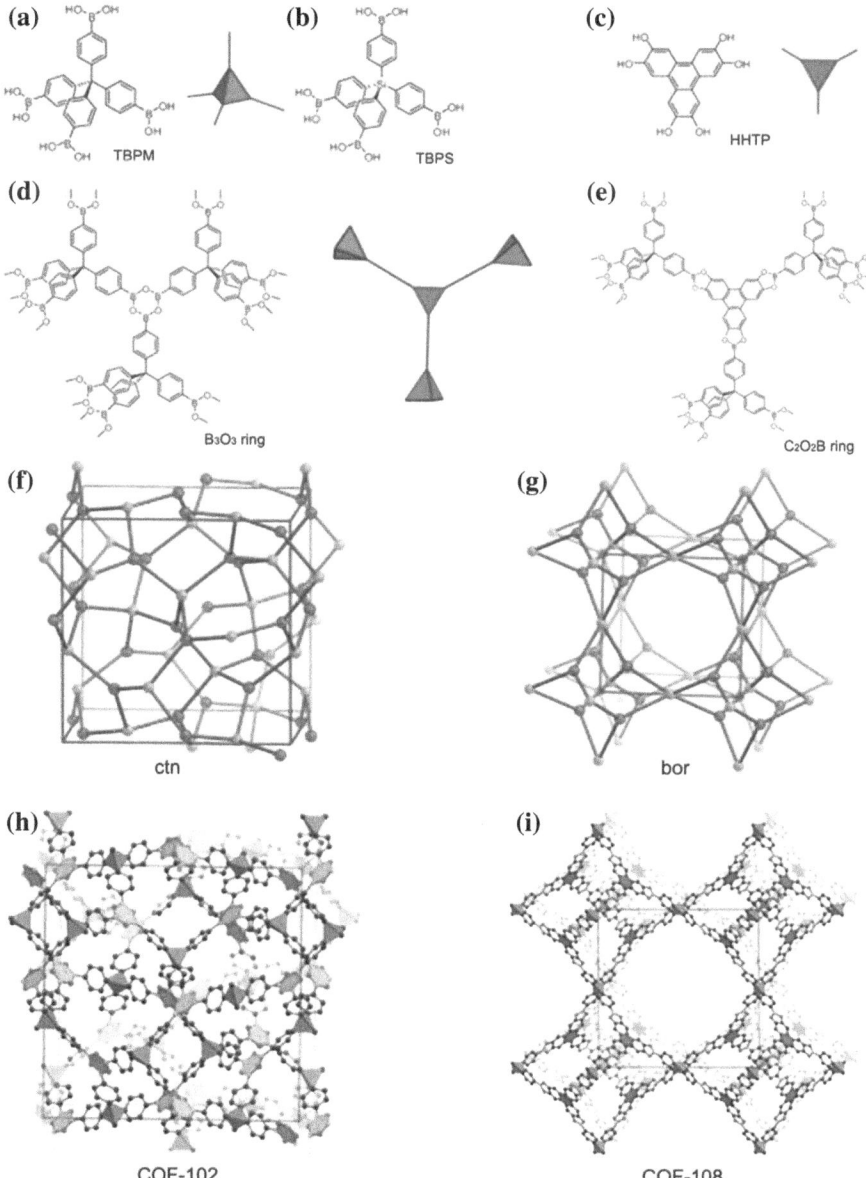

Fig. 2.15 Boronic acids are shown as tetrahedral building units in (**a**) and (**b**), and a planar tri-angular unit (**c**) is also shown (polyhedron in *orange* and triangle in *blue*, respectively) includ-ing fragments revealing the B_3O_3 (**d**) and the C_2O_2B (**e**) ring connectivity in the expected linked products via condensation routes. The represented topological nets of ctn (**f**) and bor (**g**); atomic connectivity and structure of crystalline products of COF-102 (**h**) and COF-108 (**i**). Carbon, boron, and oxygen atoms are represented as *gray*, *orange*, and *red spheres*, respectively. Adapted with permission from Ref. [11]. Copyright 2007 AAAS

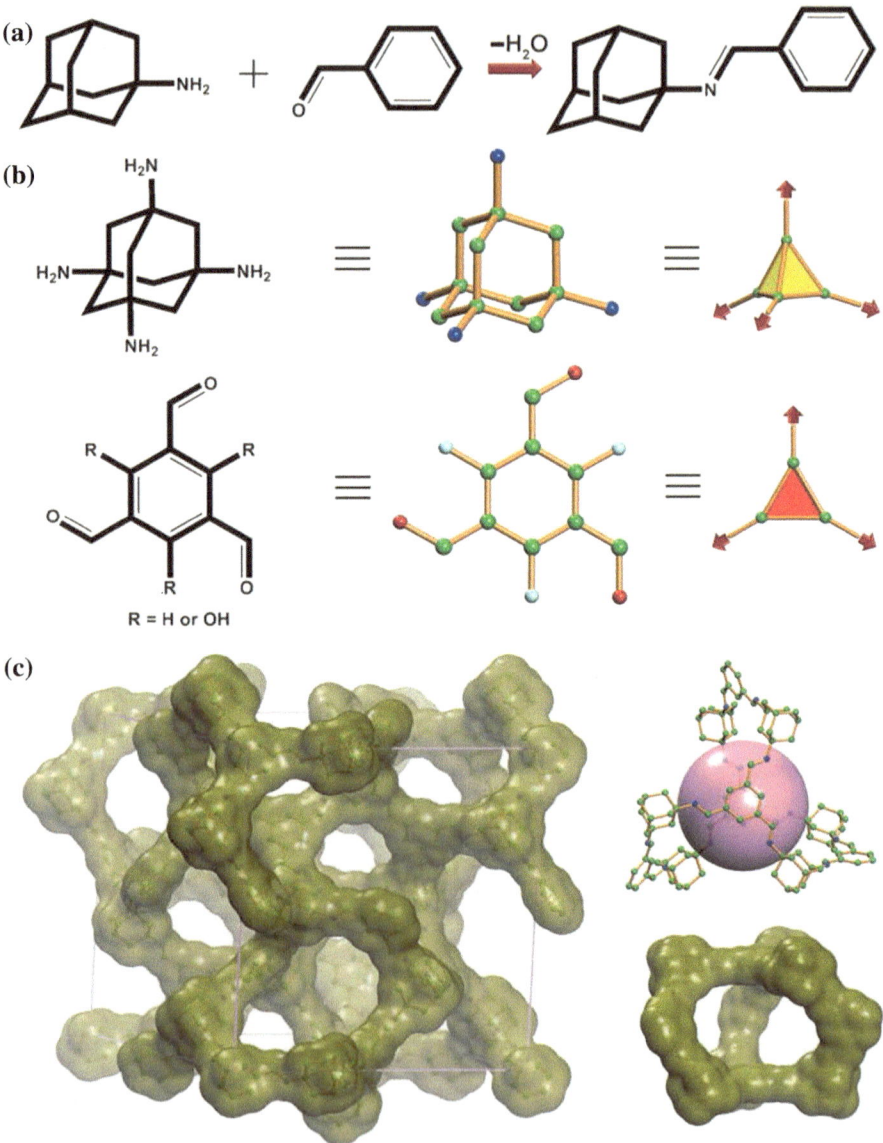

Fig. 2.16 Schematic representation of the strategy for preparing 3D microporous base-function-alized COFs. **a** Model reaction of 1-adamantanamine with benzaldehyde to form the molecule of N-(1-adamantyl) benzaldehyde imine; **b** 1,3,5,7-tetraaminoadamantane (TAA) as a tetrahedral building unit and 1,3,5-triformylbenzene (R=H, TFB) or triformylphloroglucinol (R=OH, TFP) as a triangular; **c** Extended structure of BF-COF-1. Reprinted with permission from Ref. [12]. Copyright 2014, Wiley-VCH

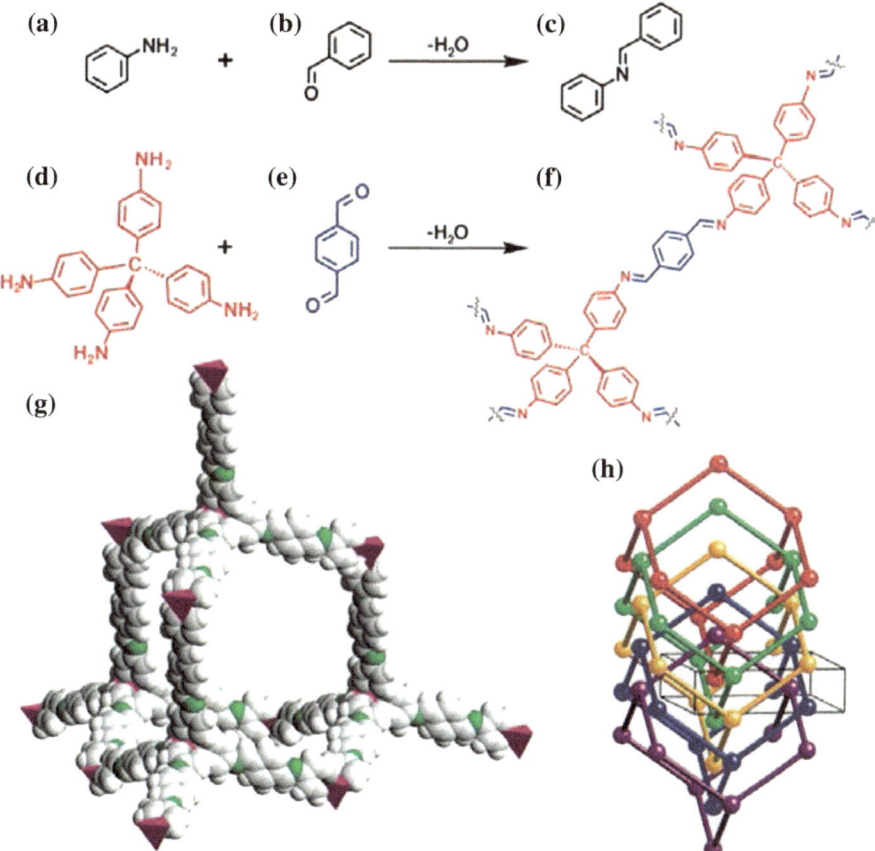

Fig. 2.17 Condensation of aniline (**a**) with benzaldehyde (**b**) forms the molecular N-benzylidene-aniline (**c**). Condensation of divergent (**d**) with ditopic (**e**) leads to the rod-like bis-imines (**f**) which will join together the tetrahedral building blocks to give the diamond structure of COF-300: **g** single framework (space filling, C *gray* and *pink*, N *green*, H *white*) and H representation of the dia-c5 topology. Reprinted with permission from Ref. [13]. Copyright 2009, American Chemical Society

an extended 3-D framework structure. Through analyzing the simulated powder patterns and the experimental measurements, COF-300 obtains a perfect match in peak position from the dia-c5 topology.

2.3.5 *Structural Studies of Crystalline POFs*

PXRD is an important tool to reveal the regularity porous materials. Deduced from XRD data, the COFs using DCC reactions are highly crystalline polymers through analyzing the simulated powder patterns and the experimental results. The structural simulation of COFs is performed by computational method according to the topology of original monomers and the reactions occur. In addition, infrared spectroscopy, solid-state NMR spectroscopy, elemental analysis, and X-ray photoelectron spectroscopy (XPS) are all useful to evaluate the linkages, terminal groups, and compositions of the COFs.

To summarize the development of different COFs, they have some common features. (1) the structures of COFs are from simple to complex due to their corresponding monomers; (2) pore sizes could be controlled; (3) the surface of COFs is adjusted by suitable function [14]. The design and synthesis of COFs have two key issues that must be considered to achieve thermodynamic control in reversible reactions: the first is the structure of the building blocks and the second is the synthetic method, including the reaction media and reaction conditions.

2.4 Amorphous POFs

Besides crystalline POFs (also referred as COFs), another major class of POFs is amorphous networks [15, 16]. Compared with crystalline POFs, the amorphous POFs are more diverse: (1) the reactions for the synthesis of amorphous POFs are more abundant than that of synthesis of crystalline POFs; (2) the active groups of monomers for the synthesis of amorphous POFs are relatively varied; (3) the structure of building blocks for the synthesis of crystalline POFs is restricted.

Amorphous POFs play vital roles in porous materials and are being actively pursued as a useful platform for advanced functional material design. Many chemists and materials scientists focus on this field and contribute to the design and synthesis of POFs. The representative POFs, including polymers of intrinsicmicroporosity (PIMs) [2–4], HCPs [17, 18], conjugated microporous polymers (CMPs) [19, 20], and porous aromatic frameworks (PAFs) [21–26], etc., constitute typical classes of covalently linked amorphous organic porous materials.

2.4.1 Polymeric Reactions

To create porosity in POFs, the key points of the current method are the choice of suitable building units and reaction medium. Chemical reactions are exploited and attempted to prepare POFs. As summarized in Fig. 2.18, several pioneering and effective strategies have been employed to construct POFs. To date, more than 20 effective reactions have been well established; representative examples t include dibenzodioxane-forming reaction [2–4], Sonogashira–Hagihara cross-coupling reaction [27–30], Suzuki cross-coupling reaction [31–33], Yamamoto type Ullmann cross-coupling reaction [34–37] and trimerization reaction of aromatic nitrile compounds [38–41], oxidative coupling reaction [42], Schiff base reaction [43], Friedel–Crafts reaction [44], etc. [45–50].

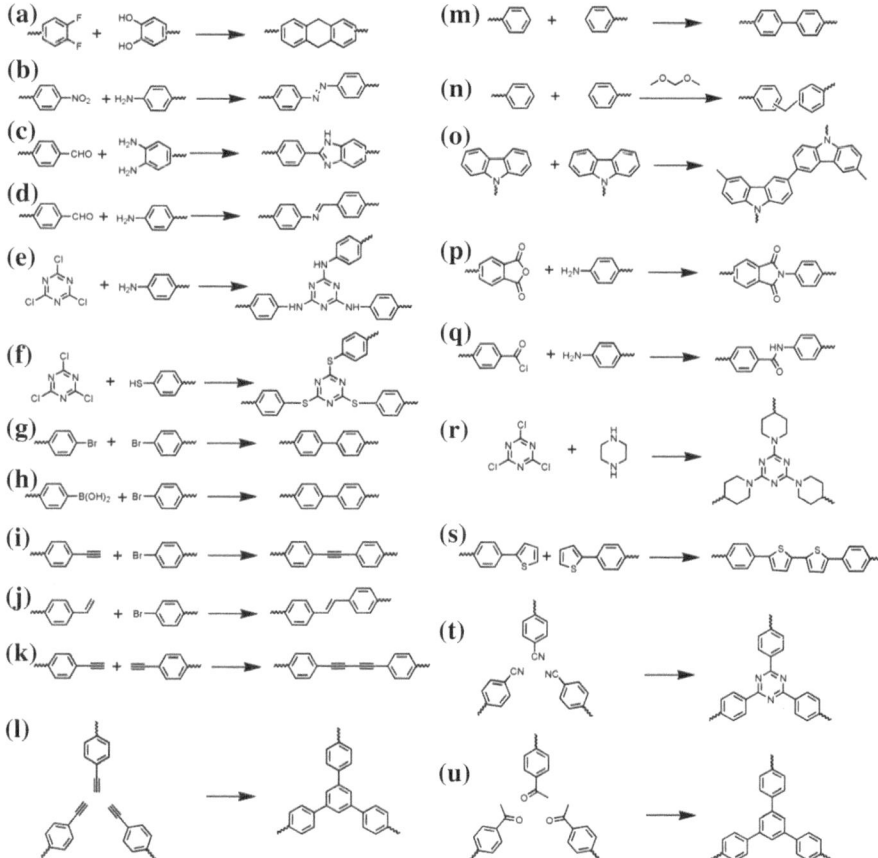

Fig. 2.18 Schematic representation of reactions for the synthesis of POFs. **a** Base-mediated aromatic nucleophilic substitution; **b** base-mediated azo formation; **c** benzimidazole formation; **d** imine formation; **e**, **f**, and **r** base-mediated nucleophilic substitution; **g** Ni(cod)$_2$-mediated aryl–aryl (Yamamoto) coupling; **h** Pd-mediated aryl–aryl (Suzuki) coupling; **i** Pd-mediated aryl–ethylynyl (Sonogashira–Hagihara) coupling; **j** Pd-mediated aryl–vinyl Heck coupliong; **k** Cu/Pd-mediated ethylynyl coupling; **l** Co-mediated ethylynyl cyclotrimerization; **m** AlCl$_3$-mediated Scholl reaction; **n** FeCl$_3$-mediated Friedel–Crafts reaction; **o** FeCl$_3$-mediated Carbazole coupling; **p** imide formation; **q** amide formation; **s** FeCl$_3$-mediated thiophenyl–thiophenyl oxidative coupling; **t** ZnCl$_2$ or super acid mediated nitrile cyclotrimerization; **u** SOCl$_2$-catalyzed acetophenone cyclotrimerization

With regard to the polymeric reactions, several issues attract great attention: (1) catalysts, such as noble metal, transition metal, and acid catalysis, and even catalyst-free; (2) reactive groups of building blocks, one reactive group means it is homocoupling reaction and more than one reactive groups implies it is cross-coupling reaction; (3) reacting conditions, including the solvents, concentration, reaction temperature, time, and reaction atmosphere. It is worth mentioning that using the imine formation reaction could obtain crystalline or amorphous POFs through controlling the reaction conditions.

Based on these effective reactions, the reactive groups of the building units are bromoarenes, iodoarenes, aromatic boronic acids, cyano-substituted arenes, aromatic aldehydes, ethynyl-substituted arenes, and amino-substituted arenes, etc. Especially, regarding AlCl$_3$-mediated Scholl reaction, the phenyl rings could directly couple together to form polymer, so the monomers do not contain reactive groups.

2.4.2 Geometric Requirements

With regard to crystalline POFs, the building units should be regular to satisfy the requirements. For amorphous POFs, the building units do not demand a regular structure. However, to build POFs with porous structure, it is necessary that the building blocks have different geometries. Using cross-coupling or self-condensation reactions of corresponding monomers, we can obtain a series of polymeric skeletons. Notably, targeted synthesis of POFs with predicted structures is extremely difficult and remains a big challenge. In the field of amorphous POFs, although they are irregular structures, the local skeleton and nanoscale of the

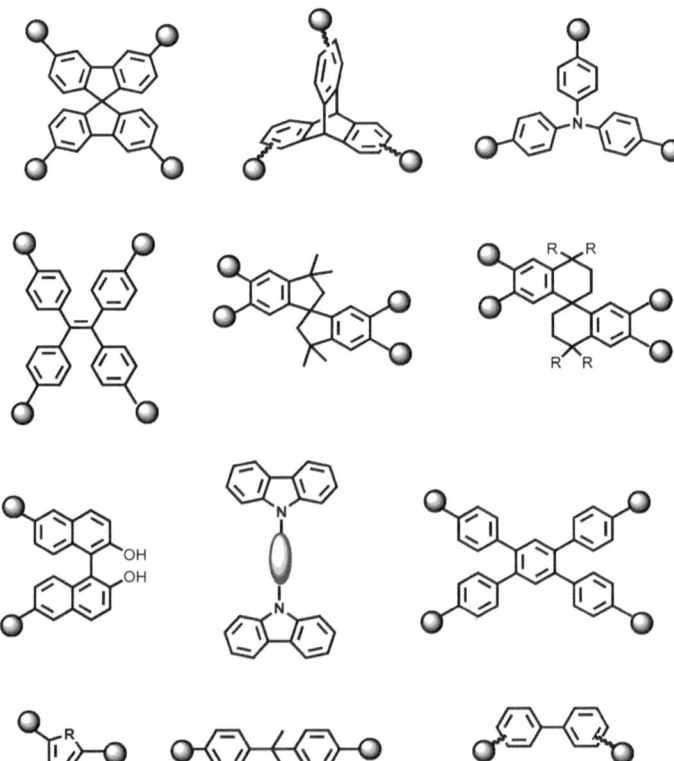

Fig. 2.19 Typical monomers with irregular configurations for the synthesis of amorphous POFs, the reactive groups could be tuned according to the polymerization reactions

products could be controlled by the initial monomers. In other words, the geometries of building blocks also play vital roles in influencing the structure of POFs. The monomers listed in Fig. 2.5 provides suitable configuration of building units to satisfy the demand on the construction of porous POFs. Because of the texture of amorphous POFs, the building units could be irregular but rigid. Figure 2.19 lists a series of monomers that could meet the requirements. It has been documented that these monomers are very powerful to construct POFs.

2.4.3 Building Units

According to the polymerization reaction, the reactive groups of building units are fixed on $-B(OH)_2$, $-Br$, $-C\equiv C$, $-C=C$, $-CHO$, $-NH_2$, and $-CN$, etc. Based on the number of reactive sites, building blocks are classified by their geometries into C2, C3, C4, and C6 categories. Generally speaking, the building units should be able to provide the suitable reactive groups and required geometries. To date, hundreds of monomers have been utilized for the construction of amorphous POFs. It is extremely difficult to list the monomers completely. Combining the geometries with the reactive groups, we can design the suitable building units. Meanwhile, the possibility of the preparation of monomers should be considered.

Figure 2.20 illustrates the typical monomers containing reactive $-Br$ group. The monomers display some clear characteristics: (1) the length of monomers is from short to long; (2) the geometries of monomers range from C2 to C3, C4, and C6, etc.; (3) the monomers are easily adjusted by adding additional groups. To achieve Sonogashira–Hagihara coupling reactions, the monomer must have the reactive $-C\equiv CH$ groups, and some typical monomers are listed in Fig. 2.21. Compared with the monomers with $-Br$ group, the monomers containing the $-C\equiv CH$ groups are rare owing to their relatively complicated synthesis procedure. Therefore, when we design the synthesis of POFs, the monomers and the reactions should be considered together.

Besides the Sonogashira–Hagihara coupling reaction, these monomers containing reactive $-Br$ group could be employed for the synthesis of amorphous POFs using aryl–aryl Yamamoto coupling, aryl–aryl Suzuki coupling, and Heck coupling reactions [51, 52]. The diversity of reactions endows POFs with different chemical structures. In addition, building units can have different geometries and diverse reactive groups, thus this structural variability significantly enhances the flexibility of the designable synthesis of POFs.

2.4.4 Characterizations of Amorphous POFs

PXRD is a powerful method for evaluating the crystallinity of POFs. For POF materials, it seems difficult to form an ideal proposed structure and an ordered connection among the building unit in the actual framework. The lack of

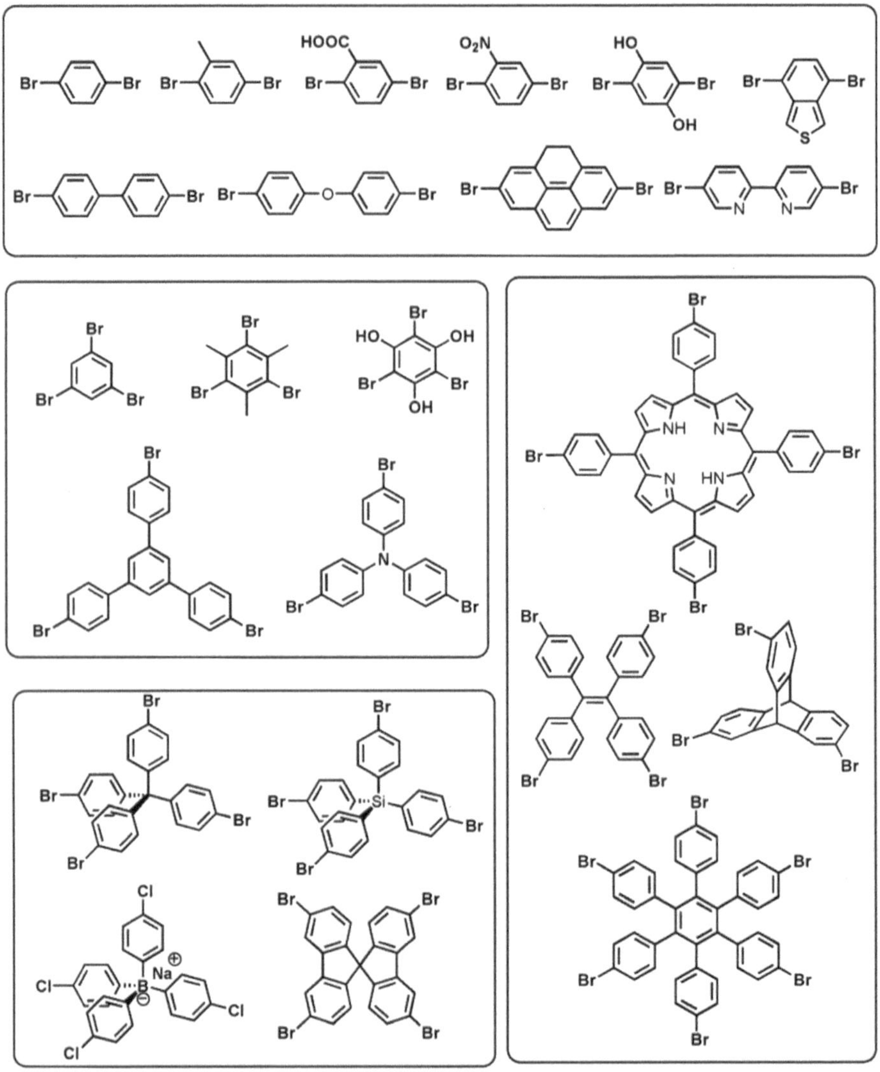

Fig. 2.20 Typical monomers with reactive –Br group for the synthesis of amorphous POFs using aryl–aryl Yamamoto coupling, aryl–aryl Suzuki coupling, Sonogashira–Hagihara coupling, Heck coupling reactions

long-ordered range structure of POFs is due to the distortion of the constructing units and kinetics controlled irreversible coupling process. Compared with soluble polymers and crystalline organic porous frameworks, the structural characterizations of POFs are relatively complicated due to their insolubility and amorphous nature. It requires much effort to unambiguously characterize the structure of POFs by the combination of various chemical analytical methods.

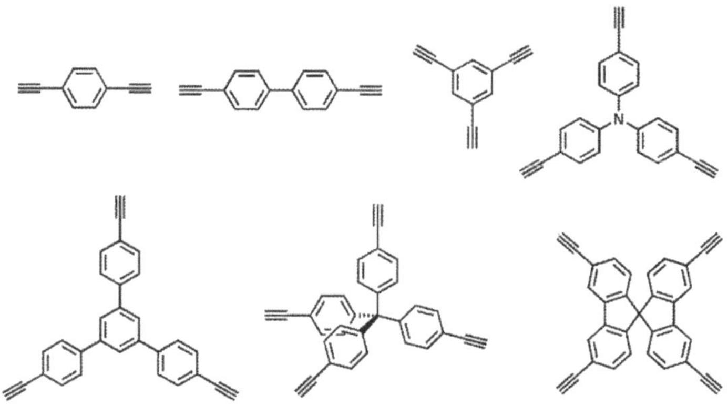

Fig. 2.21 Typical monomers with reactive –C≡CH group for the synthesis of amorphous POFs

For example, infrared spectroscopy is carried out to detect the success and completion of reaction indicated by the change in the characteristic band. Elemental analysis is powerful to analyze the composition of POFs. In addition, XPS is intensive to detect the elements except for H. Therefore, XPS could be utilized to confirm the ending groups of POFs and the residual of catalysis. Solid-state ^1H and ^{13}C nuclear magnetic resonance (NMR) spectroscopy is performed to investigate the local structure of POFs. The macroscopic morphology of POFs is studied through field emission scanning electron microscopy for evaluating the size and morphology. High-resolution transmission electron microscopy could investigate their porous texture. To study the 3D structure, Cooper and coworkers have developed an atomistic simulation method [53]. The method utilizes fragmental models and amorphous cell simulation embedded in the Materials Studio Modeling package to build the molecular models. Nitrogen adsorption and desorption isotherms are useful to provide information concerning the porosity of POPs, including some important parameters such as surface area, total pore volume, microporous pore volume, mesoporous pore volume, and pore size distribution.

2.5 Concerns in the Development of POFs

POFs as an important subclass of nanoporous materials are of great interest in materials science. In recent years, the discovery of POFs for advanced applications has attracted extensive attention and intensive efforts. In fact, POF materials have contributed to various fields, potentially for use in the areas of gas storage, molecular separations, sensors, catalysts [54, 55], etc. As a result, the design of materials with multifunctionalities is an ever-pursued dream of materials scientists and engineers. In the development of POF materials, some important issues are of great interest.

2.5.1 Initial Stage of the Development of POFs

Before 2005, there were few reports concerning porous polymers. McKeown and Budd et al. reported a series of PIMs [2–4]. In 2005, Yaghi et al. first reported 2D crystalline polymers, COF-1 and COF-5 [5]. Subsequently, a series of 3D crystalline polymers were successfully designed and synthesized by the same group in 2007 [11]. Meanwhile, Cooper et al. illustrated the synthesis of CMPs [27]. At the beginning, few scientists successfully prepared and reported the synthesis of POFs, and monomers and reactions were relatively rare. Because POFs display so many eye-catching properties, our group paid close attention to this field at the beginning of 2007. However, it is difficult to achieve this goal because the synthesis and characterization of POFs is very complicated. In 2006, we followed the interest in the design and synthesis of POFs. Until 2009, we reported PAF-1 with exceptionally high surface area, which are employed by aryl–aryl Yamamoto coupling reaction [56]. To obtain POFs with excellent properties, we need continuous exploration. Based on our research process, it is estimated that many groups had focused on this field at that time. After 2009, many strategies were exploited and utilized for the synthesis of POFs, which are successfully employed to prepare linear traditional polymers, such as Sonogashira–Hagihara coupling, aryl–aryl Yamamoto coupling, and aryl–aryl Suzuki coupling, etc.

2.5.2 Development of New Reactions

In studies of POFs, we could meet some problems, such as their stability, cost, etc. B_3O_3 ring formation and BO_2C_2 ring formation are powerful to construct COFs. However, the resulting COFs show inferior stability when they suffer from moisture, even in the open atmosphere. With regard to palladium and nickel (0) compounds as catalyst, they are expensive and oxygen sensitive. In addition, the corresponding monomers are strictly limited those having special reactive groups, such as –Br, –B(OH)$_2$, –C≡CH, and –C≡CH$_2$, which are relatively difficult to prepare. Therefore, they should face one or more of the following difficulties: (1) requirement of drastic synthesis conditions; (2) tedious purification processes; (3) difficulties in reproducing results; (4) multistep synthesis from the aromatic hydrocarbon precursor; (5) the formation of complicated mixtures due to competing reactions. In other words, the traditional methods of constructing POFs are not only complicated but also expensive, thereby making them difficult in the case of large-scale production.

Seeking relatively inexpensive catalyst and easily obtained monomers will assist the development of POFs. One successful example is a new strategy reported by Tan et al. [17, 57, 58], which is knitting rigid aromatic building blocks by external cross-linker. Another is the Scholl coupling reaction with AlCl$_3$ as the catalyst. Our group and Tan et al. had used this reaction for synthesis of POFs almost at the same time [59–61]. The coupling reaction could occur between the phenyl rings of aromatic compounds (Fig. 2.22). It is also worth mentioning that the developed approach also overcomes typical flaws of some classic POFs, such as high cost and complexity of precursor preparation.

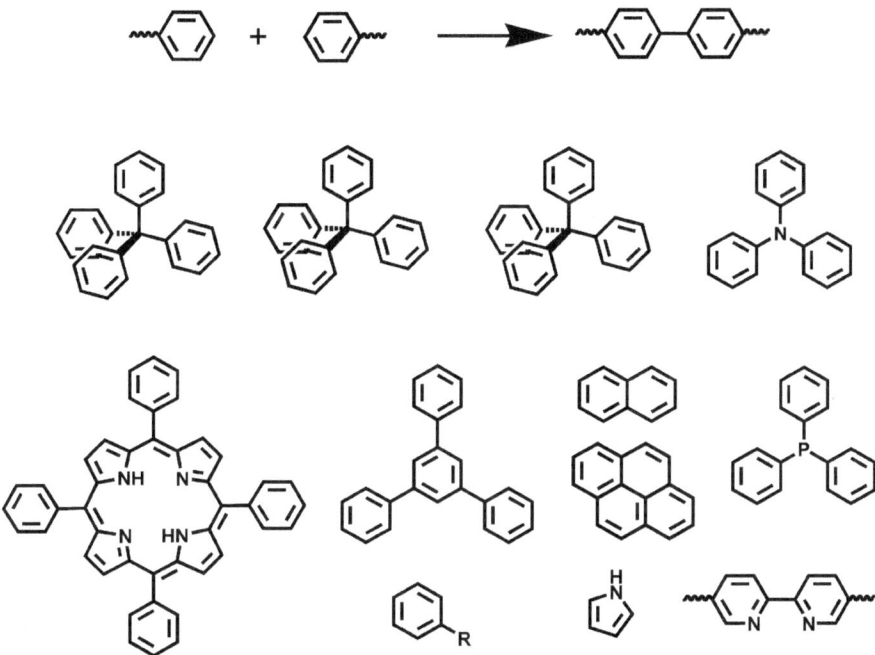

Fig. 2.22 Scholl coupling reaction to obtain extended frameworks, and the used monomers

2.5.3   Controlling of Reaction Conditions

When polymerization reaction occurs, many factors would affect the property of final product, including reaction media, reaction temperature, type of catalyst, concentration, reaction time, etc.

As we know, the solvents used for Sonogashira–Hagihara cross-coupling reaction mainly include THF, DMF, 1,4-dioxane and toluene systems. For example, the copolymerization of 1,3,5-triethynylbenzene with 2,5-dibromobenzene-1,4-diol are synthesized using above-mentioned solvents, whose surface areas are 847, 1,043, 778, and 761 m^2 g^{-1}, respectively [62]. However, it is still not clear which solvent is the best choice for Sonogashira–Hagihara cross-coupling reaction. We also screen the solvent effect using Suzuki cross-coupling reaction. PAF-12 [63], derived from octaphenylsilsesquioxane as basic building units, are obtained using different solvent systems, DMF/H_2O, 1,4-dioxane/H_2O, and THF/H_2O. Although the surface areas of the materials are not of much difference, in DMF/H_2O system no iodine end groups were detected by XPS (the average molar ratio of I/Si < 0.001). In contrast, I/Si molar ratios of samples synthesized in other two solvent systems are 0.097 (1,4-dioxane/H_2O) and 0.295 (THF/H_2O), respectively. These results indicate that the reaction in DMF/H_2O system proceeds more completely.

In the ionothermal synthesis of triazine-based porous polymers, the ratio of catalyst to monomer affects the polymerization degree significantly, giving rise to a series of networks with different pore volumes, surface areas, and pore size distributions. For example, the porous triazine network PAF-16 was synthesized from a mixture of tetrakis(4-cyanophenyl)silica and anhydrous $ZnCl_2$ with different proportions at 400 °C [41]. The use of 1-equivalent $ZnCl_2$ leads to the formation of a polymer with a surface area of 190 m^2 g^{-1}, which increases to 979 and 1,900 m^2 g^{-1} when a 5-equivalent catalyst and 10-equivalent catalyst are employed. Additionally, in the ionothermal synthesis of triazine-based porous polymers, the reaction temperature also plays an important role in affecting the porosity. For example, trimerization of 1,4-dicyanobenzene is performed at different temperatures (400, 500, 600, and 700 °C), the BET surface area increases from 920 to 1,600, 1,750, and 2,530 m^2 g^{-1}, respectively [8].

As a basic synthetic method, the Suzuki coupling reaction was used for the production of four polymeric frameworks (Fig. 2.23), referred to as EOF-6, EOF-7,

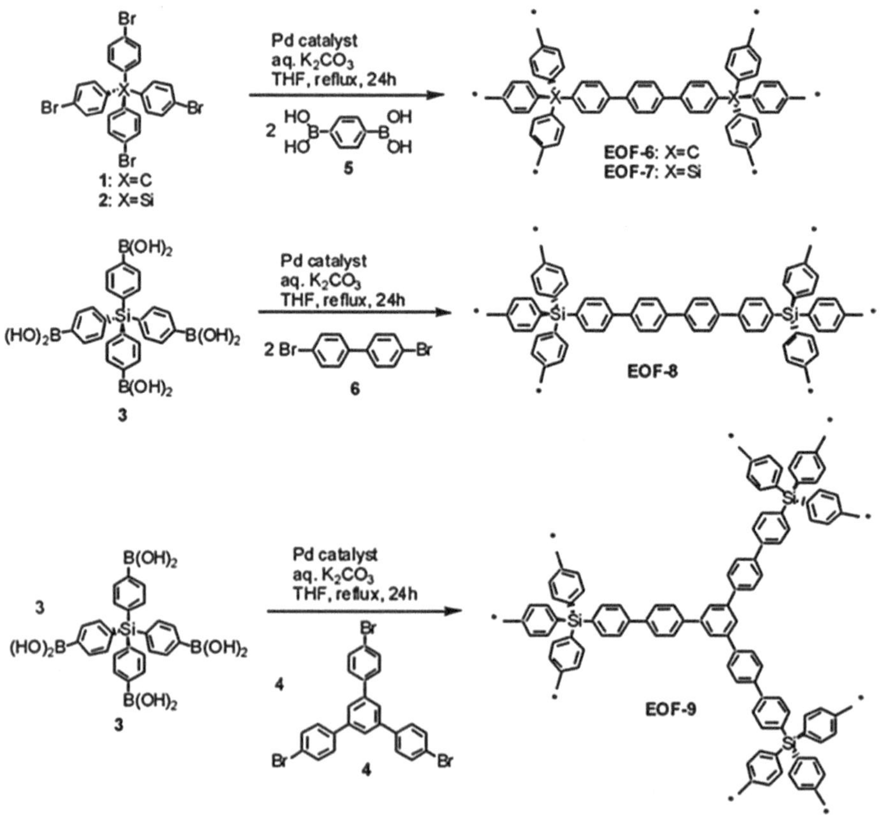

Fig. 2.23 Syntheses of EOF-6 to EOF-9. Reprinted with permission from Ref. [64]. Copyright 2010, Royal Society of Chemistry

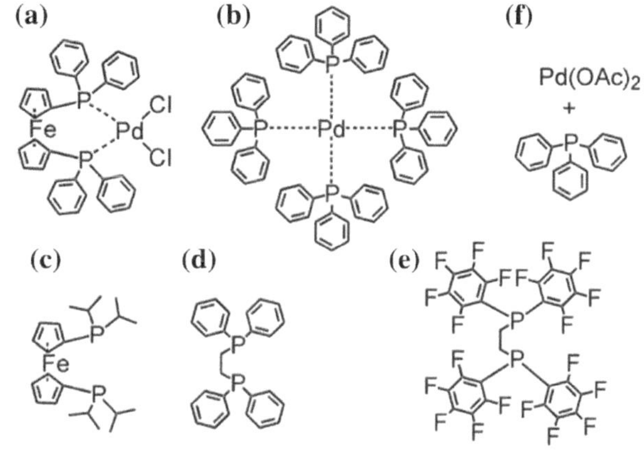

Fig. 2.24 Pd catalysts (**a, b** and **f**) and ligands (**c–e**) used for Suzuki coupling reaction. Reprinted with permission from Ref. [64]. Copyright 2010, Royal Society of Chemistry

Table 2.1 Dependence of the specific surface area of EOF-6 on different catalysts and solvents used in the synthesis

Catalyst	Solvent	$SSA^a/m^2\,g^{-1}$
a	Tetrahydrofuran	835
a	Tetrahydrofuran	990
a	Diethyl ether	No polymer
b	Tetrahydrofuran	552
a + c	Tetrahydrofuran	No polymer
b + c	Tetrahydrofuran	1,380
b + d	Tetrahydrofuran	827
b + e	Tetrahydrofuran	No polymer
a + c	Dioxane	1,133
a + c	1,2-Dimethoxy ethane	364
f	*N,N*-Dimethyl formamide	46

Reprinted with permission from Ref. [64]. Copyright 2010, Royal Society of Chemistry
[a]Specific surface area determined as single point BET from N_2 physisorption isotherm at $P/P_o = 0.3$

EOF-8, and EOF-9 [64]. The experiments were performed with different catalytic systems (Fig. 2.24), which intensively affected the final product (Table 2.1).

2.5.4 Development of Applications of Porous Materials

Porous materials have many applications due to their porosity. With regard to POFs, they are commonly used as gas storage, and most reports focus on their H_2, CO_2, and CH_4 sorption capacities. On the basis of their self-structural characteristic, POFs show glamorous applications in various fields, including sensor, catalysis, host–guest chemistry, etc., and their advanced applications will be discussed in detail in the following chapters.

References

1. Zou X, Ren H, Zhu G (2013) Topology-directed design of porous organic frameworks and their advanced applications. Chem Commun 49:3925–3936
2. McKeown N, Makhseed S, Budd P (2002) Phthalocyanine-based nanoporous network polymers. Chem Commun 2780–2781
3. McKeown N, Hanif S, Msayib K et al (2002) Porphyrin-based nanoporous network polymers. Chem Commun 2782–2783
4. McKeown N, Budd P, Msayib K et al (2005) Polymers of Intrinsic Microporosity (PIMs): bridging the void between microporous and polymeric materials. Chem Eur J 11:2610–2620
5. Cote A, Benin A, Ockwig N et al (2005) Porous, crystalline, covalent organic frameworks. Science 310:1166–1170
6. Ding S, Wang W (2013) Covalent organic frameworks (COFs): from design to applications. Chem Soc Rev 42:548–567
7. Feng X, Ding X, Jiang D (2012) Covalent organic frameworks. Chem Soc Rev 41:6010–6022
8. Kuhn P, Antonietti M, Thomas A (2008) Porous covalent triazine-based frameworks prepared by ionothermal synthesis. Angew Chem Int Ed 47:3450–3453
9. Tilford R, Mugavero S, Pellechia P (2008) Tailoring microporosity in covalent organic frameworks. Adv Mater 20:2741–2746
10. Spitler E, Colson J, Uribe-Romo F et al (2012) Lattice expansion of highly oriented 2D phthalocyanine covalent organic framework films. Angew Chem Int Ed 51:2623–2627
11. El-Kaderi H, Hunt J, Mendoza-Cortes J et al (2007) Designed synthesis of 3D covalent organic frameworks. Science 316:268–272
12. Fang Q, Gu S, Zheng J et al (2014) 3D Microporous base-functionalized covalent organic frameworks for size-selective catalysis. Angew Chem Int Ed 53:2878–2882
13. Uribe-Romo F, Hunt J, Furukawa H et al (2009) A crystalline imine-linked 3-D porous covalent framework. J Am Chem Soc 131:4570–4571
14. Nagai A, Guo Z, Feng X et al (2011) Pore surface engineering in covalent organic frameworks. Nat Commun 2:536–543
15. McKeown N, Budd P (2010) Exploitation of intrinsic microporosity in polymer-based materials. Macromolecules 43:5163–5176
16. Dawson R, Cooper A, Adams D (2012) Nanoporous organic polymer networks. Prog Polym Sci 37:530–563
17. Li B, Gong R, Wang W et al (2011) A new strategy to microporous polymers: Knitting rigid aromatic building blocks by external cross-linker. Macromolecules 44:2410–2414
18. Xu S, Luo Y, Tan B (2013) Recent development of hypercrosslinked microporous organic polymers. Macromol Rapid Commun 34:471–484
19. Vilela F, Zhang K, Antonietti M (2012) Conjugated porous polymers for energy applications. Energy Environ Sci 5:7819–7832
20. Xu Y, Jin S, Xu H et al (2013) Conjugated microporous polymers: design, synthesis and application. Chem Soc Rev 42:8012–8031
21. Yuan Y, Ren H, Sun F et al (2012) Sensitive detection of hazardous explosives via highly fluorescent crystalline porous aromatic frameworks. J Mater Chem 22:24558–24562
22. Yuan Y, Sun F, Ren H et al (2011) Targeted synthesis of a porous aromatic framework with a high adsorption capacity for organic molecules. J Mater Chem 21:13498–13502
23. Ma H, Ren H, Meng S et al (2013) A 3D microporous covalent organic framework with exceedingly high C_3H_8/CH_4 and C_2 hydrocarbon/CH_4 selectivity. Chem Commun 49:9773–9775
24. Ma H, Ren H, Zou X et al (2014) Post-metalation of porous aromatic frameworks for highly efficient carbon capture from $CO_2 + N_2$ and $CH_4 + N_2$ mixtures. Polym Chem 5:144–152
25. Yuan R, Ren H, Yan Z et al (2014) Robust tri(4-ethynylphenyl)amine-based porous aromatic frameworks for carbon dioxide capture. Polym Chem 5:2266–2272

26. Ma H, Ren H, Meng S et al (2013) Novel porphyrinic porous organic frameworks for high performance separation of small hydrocarbons. Sci Rep 3:267–2611
27. Jiang J, Su F, Trewin A et al (2007) Conjugated microporous poly(aryleneethynylene) networks. Angew Chem Int Ed 46:8574–8578
28. Jiang J, Trewin A, Su F et al (2009) Microporous poly(tri(4-ethynylphenyl)amine) networks: synthesis, properties and atomistic simulation. Macromolecules 42:2658–2666
29. Yan J, Ren H, Ma H et al (2013) Construction and sorption properties of pyrene-based porous aromatic frameworks. Micro Meso Mater 173:92–98
30. Wang Z, Yuan S, Mason A et al (2012) Nanoporous porphyrin polymers for gas storage and separation. Macromolecules 45:7413–7419
31. Xu Y, Nagai A, Jiang D (2013) Core–shell conjugated microporous polymers: a new strategy for exploring color-tunable and -controllable light emissions. Chem Commn 49:1591–1593
32. Zhang L, Lin T, Pan X et al (2012) Morphology-controlled synthesis of porous polymer nanospheres for gas absorption and bioimaging applications. J Mater Chem 22:9861–9869
33. Jeon H, Choi J, Lee Y et al (2012) Highly selective CO_2-capturing polymeric organic network structures. Adv Energy Mater 2:225–228
34. Zhang Q, Zhang S, Li S (2012) Novel functional organic network containing quaternary phosphonium and tertiary phosphorus. Macromolecules 45:2981–2988
35. Xiang Z, Cao D (2012) Synthesis of luminescent covalent-organic polymers for detecting nitroaromatic explosives and small organic molecules. Macromol Rapid Commun 33:1184–1190
36. Ren H, Ben T, Sun F et al (2011) Synthesis of a porous aromatic framework for adsorbing organic pollutants application. J Mater Chem 21:10348–10353
37. Kassab R, Jackson K, El-Kadri O et al (2011) Nickel-catalyzed synthesis of nanoporous organic frameworks and their potential use in gas storage applications. Res Chem Intermed 37:747–757
38. Kuhn P, Forget A, Su D et al (2008) From microporous regular frameworks to mesoporous materials with ultrahigh surface area: dynamic reorganization of porous polymer networks. J Am Chem Soc 130:13333–13337
39. Kuhn P, Thomas A, Antonietti M (2009) Toward tailorable porous organic polymer networks: a high-temperature dynamic polymerization scheme based on aromatic nitriles. Macromolecules 42:319–326
40. Hug S, Tauchert M, Li S et al (2012) A functional triazine framework based on N-heterocyclic building blocks. J Mater Chem 22:13956–13964
41. Wang W, Ren H, Sun F et al (2012) Synthesis of porous aromatic framework with tuning porosity via ionothermal reaction. Dalton Trans 41:3933–3936
42. Chen Q, Luo M, Hammershøj P et al (2012) Microporous polycarbazole with high specific surface area for gas. J Am Chem Soc 134:6084–6087
43. Zhu Y, Long H, Zhang W (2013) Imine-linked porous polymer frameworks with high small gas (H_2, CO_2, CH_4, C_2H_2) uptake and CO_2/N_2 selectivity. Chem Mater 25:1630–1635
44. Chaikittisilp W, Kubo M, Moteki T et al (2011) Porous siloxane organic hybrid with ultrahigh surface area through simultaneous polymerization destruction of functionalized cubic siloxane cages. J Am Chem Soc 133:13832–13835
45. Rabbani M, El-Kaderi H (2012) Synthesis and characterization of porous benzimidazole-linked polymers and their performance in small gas storage and selective uptake. Chem Mater 24:1511–1517
46. Jiang M, Wang Q, Chen Q et al (2013) Preparation and gas uptake of microporous organic polymers based on binaphthalene-containing spirocyclic tetraether. Polymer 54:2952–2957
47. Yuan S, Shui J, Grabstanowicz L et al (2013) A highly active and support-free oxygen reduction catalyst prepared from ultrahigh-surface-area porous polyporphyrin. Angew Chem Int Ed 52:8349–8353
48. Han Y, Zhang L, Zhao Y et al (2013) Microporous organic polymers with ketal linkages: synthesis, characterization, and gas sorption properties. ACS Appl Mater Interf 5:4166–4172
49. Tanabe K, Siladke N, Broderick E et al (2013) Stabilizing unstable species through single-site isolation: a catalytically active TaV trialkyl in a porous organic polymer. Chem Sci 4:2483–2489

50. Lieb M, Senker J (2013) Microporous functionalized triazine-based polyimides with high CO_2 capture capacity. Chem Mater 25:970–980
51. Sun L, Zou Y, Liang Z et al (2014) A one-pot synthetic strategy via tandem Suzuki-Heck reactions for the construction of luminescent microporous organic polymers. Polym Chem 5:471–478
52. Sun L, Liang Z, Yu J et al (2013) Luminescent microporous organic polymers containing the 1,3,5-tri(4-ethenylphenyl)benzene unit constructed by Heck coupling reaction. Polym Chem 5:1932–1938
53. Jiang J, Su F, Trewin A et al (2008) Synthetic control of pore dimension and surface area in conjugated microporous polymer and copolymer networks. J Am Chem Soc 130:7710–7720
54. Kaur P, Hupp J, Nguyen S (2011) Porous organic polymers in catalysis: opportunities and challenges. ACS Catal 1:819–835
55. Zhang Y, Riduan S (2012) Functional porous organic polymers for heterogeneous catalysis. Chem Soc Rev 41:2083–2094
56. Ben T, Ren H, Ma S et al (2009) Targeted synthesis of a porous aromatic framework with high stability and exceptionally high surface area. Angew Chem Int Ed 48:9457–9460 (Angew Chem 9621–9624)
57. Li B, Guan Z, Wang W et al (2012) Highly dispersed Pd catalyst locked in knitting aryl network polymers for Suzuki-Miyaura coupling reactions of aryl chlorides in aqueous media. Adv Mater 24:3390–3395
58. Luo Y, Li B, Wang W et al (2012) Hypercrosslinked aromatic heterocyclic microporous polymers: a new class of highly selective CO_2 capturing materials. Adv Mater 24:5703–5707
59. Meng S, Ma H, Jiang L et al (2014) A facile approach to prepare porphyrinic porous aromatic frameworks for small hydrocarbon separation. J Mater Chem A 2:14536–14541
60. Li L, Ren H, Yuan Y et al (2014) Construction and adsorption properties of porous aromatic frameworks via $AlCl_3$-triggered coupling polymerization. J Mater Chem A 2:11091–11098
61. Li B, Guan Z, Yang X et al (2014) Multifunctional microporous organic polymers. J Mater Chem A 2:11930–11939
62. Dawson R, Laybourn A, Khimyak Y et al (2010) High surface area conjugated microporous polymers: the importance of reaction solvent choice. Macromolecules 43:8524–8530
63. Jing X, Sun F, Ren H et al (2013) Targeted synthesis of micro–mesoporous hybrid material derived from octaphenylsilsesquioxane building units. Micro Meso Mater 165:92–98
64. Rose M, Klein N, B€ohlmann B et al (2010) New element organic frameworks via Suzuki coupling with high adsorption capacity for hydrophobic molecules. Soft Matter 6:3918–3923

Chapter 3
Synthetic Post-modification of Porous Organic Frameworks

Abstract POFs are constructed by organic building units via polymerization reactions. To satisfy the proposed requirements, original organic monomers could be designed and prepared using the powerful synthetic strategy of organic chemistry. Therefore, special functional groups could be attached to the POF skeletons. In this chapter, we introduce some typical and successful strategies that are utilized for synthetic post-modification of POFs. One effective strategy is the rational use of POFs containing active sites such as azide and alkyne for click chemistry, ion exchange, metallization of –OH and –COOH. Another is using the phenyl ring of POFs, which is a good precursor of the preparation of many functional groups containing compounds.

Keywords Post-modification · Reactive sites · Click reaction · Ion exchange · Metallization · Phenyl rings · Functional groups

3.1 Introduction

Nowadays, hundreds of POFs have been successfully designed and synthesized [1]. The construction of POFs is based on the building units, also called monomers. Monomers could be rationally designed and prepared to satisfy the proposed requirements owing to synthetic features of organic chemistry (Fig. 3.1). Therefore, special functional groups could be attached to the building units. As a result, POFs with reactive sites are obtained. To change the porosity of the resulting POF materials, synthetic post-modification should be performed (Fig. 3.1). In this chapter, we introduce some typical and successful strategies.

© The Author(s) 2015
G. Zhu and H. Ren, *Porous Organic Frameworks*, SpringerBriefs in Green Chemistry for Sustainability, DOI 10.1007/978-3-662-45456-5_3

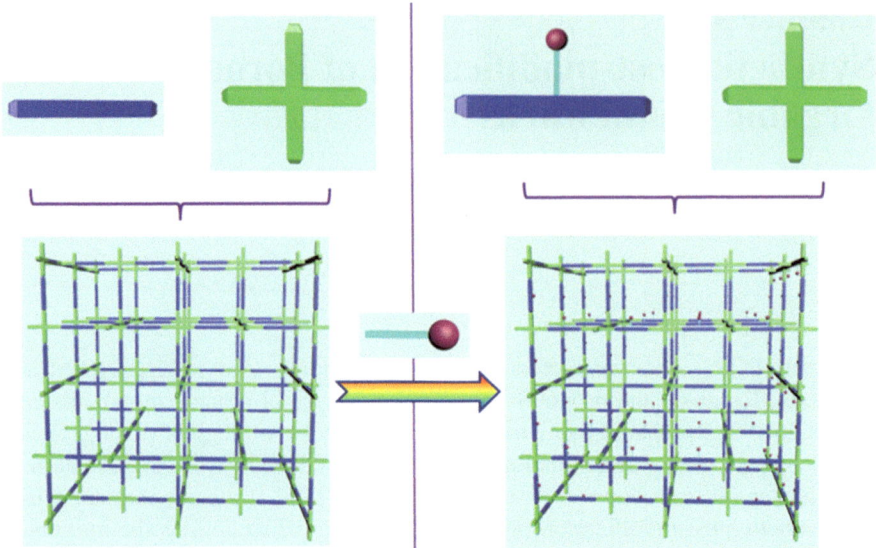

Fig. 3.1 The promising methods for constructing POFs with reactive sites via de novo or post-modification method

3.2 POFs Containing Reactive Sites

To perform synthetic post-modification of POFs, one simple and effective strategy is the introduction of functional groups into POFs to form reactive sites containing POFs. When designing the structure of POFs, we should consider what kind of groups can be easily transformed. Common reactive groups include $-NH_2$, $-CHO$, $-OH$, $-COOH$, $-N_3$, $-SH$, $-C\equiv CH$, $-NO_2$, etc. The knowledge of common organic reactions is useful for designing the transformation of functional groups. In this section, we introduce some representative examples that demonstrate successful changes in the parent material structure.

3.2.1 "Click" Chemistry

Click chemistry is a term applied to chemical synthesis tailored to generate substances quickly and reliably by joining small units together. One of the most popular reactions within the Click chemistry concept is the azide alkyne Huisgen cycloaddition using a Copper (Cu) catalyst at room temperature [2]. It has been documented that click chemistry is an effective method for construction of POFs.

The reactive groups of click reaction are azide and alkyne. Jiang's group illustrated a strategy using click reaction to tune the pore sizes of two COF materials [3]. One azide-decorated building unit (azide-appended benzenediboronic acid)

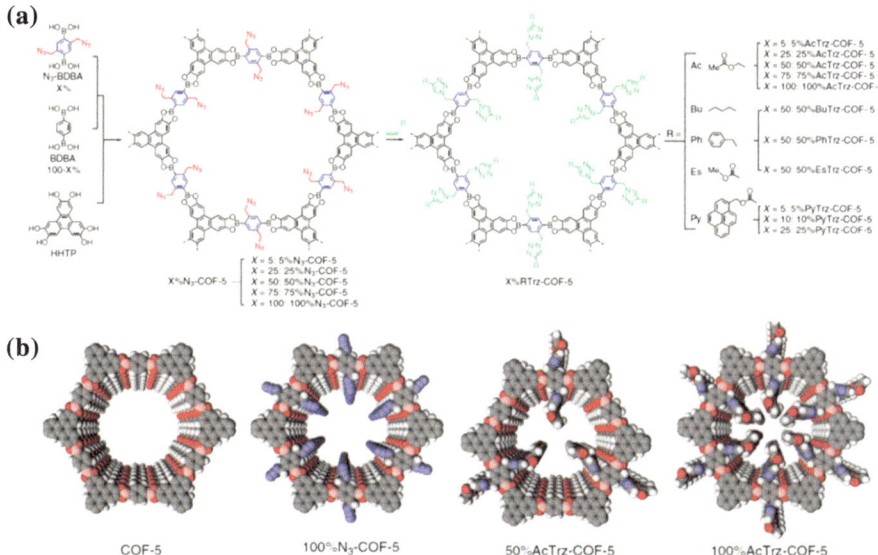

Fig. 3.2 Pore surface engineering in COFs. **a** The scheme shows a general strategy for surface engineering of COFs through the combination of condensation reaction and click chemistry. In the first step, COFs bearing azide units on the walls are synthesized by condensation reaction of HHTP with azide-appended benzene diboronic acid (N_3-BDBA) and benzene diboronic acid (BDBA) in a designated molar ratio ($X = 0$–100 %). The content of N-appended wall units is tunable from 0 to 100 %, depending on the molar ratio of N_3-BDBA to BDBA. Five members of $X\%N_3$-COF-5 with different contents ($X = 5, 25, 50, 75,$ and 100) were synthesized. The structure of $100\%N_3$-COF-5 is shown in the figure, with all of the walls occupied by the N_3-appended phenylene units. In the next step, the azide groups on the COF walls are clicked with alkynes to anchor various organic groups onto the walls of COFs ($X\%RTrz$-COF-5). The density of surface R groups on the walls is determined by the azide content in $X\%N_3$-COF-5. **b** Graphical representation of COF-5 upon surface engineering, which leads to the functionalization of organic groups on the walls. The component, composition, and density of the organic groups on the walls are controllable. Reprinted with permission from Macmillan Publishers Ltd.: Ref. [3], copyright 2011

was used in the synthesis procedures of COF-5 (Fig. 3.2) and NiPc-COF. Through controlling the azide compound content, a series of azide decorated COF-5 and NiPc-COF products were obtained. The azide groups provided a favorable opportunity for the post-synthesis of COFs via effective click reaction. As a result, introducing triazole-linked moieties into the channels of COF tuned the pore size continuously. The designer successfully decorated COF-5 and NiPc-COF with different amounts (5, 25, 50, 75 and 100 %) of azide-decorated phenyl boronic acids.

Recently, Jiang et al. reported a synthetic strategy for constructing organocatalytic covalent organic frameworks via pore surface engineering [4]. Alternatively, the alkyne group was introduced into the original materials with varied molar ratios (Fig. 3.3). Then the azide compound with pyrrolidine unit was utilized for post-modification. The catalytic activity of the resulting product depends on the density of the active sites on the pore walls.

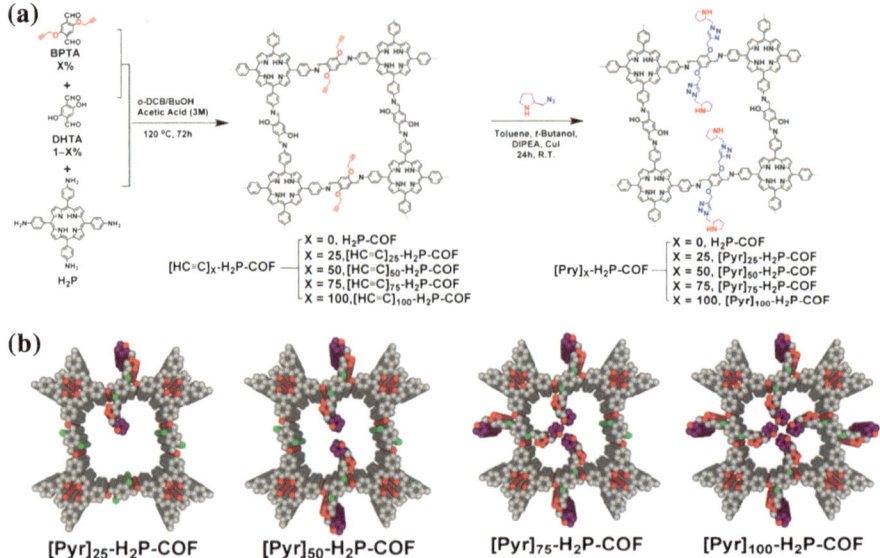

Fig. 3.3 **a** The general strategy for pore surface engineering of imine-linked COFs via condensation reaction and click chemistries (the case for $X = 50$ was exemplified); **b** a graphical representation of [Pyr]X-H₂P-COF with different densities of catalytic sites on the pore walls (*gray* carbon, *red* nitrogen, *green* oxygen, *purple* carbon atoms of the pyrrolidine unit; hydrogen is omitted for clarity). Reprinted with permission from Ref. [4]. Copyright 2014, Royal Society of Chemistry

COF-102 reported by Yaghi et al. was obtained by self-condensation of tetrahedral tetrakis(boronic acid) [5]. Allyl group presents a truncated mixed monomer approach to incorporate arbitrary functional groups within the interior of 3D COFs [6]. By co-condensation of tetrakis(boronic acid) and allyl-functionalized tris(boronic acid), a truncation–functionalization 3D COF-102 was prepared (Fig. 3.4a). The allyl groups incorporated within the pore walls were subjected to thiol–ene coupling conditions (Fig. 3.4b). The resulting product also maintained the crystallinity and permanent porosity of the parent framework.

3.2.2 Ion Exchange

As we know, zeolites possess charged skeletons. There are different cations, such as metal ions, occupying the inner wall of the pores of zeolites. Therefore, the pore sizes could be tuned by the types of the extra-framework metal ions. For example, the pore size of zeolite A could be tuned from 3 to 5 Å using K⁺, Na⁺, and Ca²⁺ (Fig. 3.5). The ion exchanging strategy is also successfully employed to adjust the structure of POFs.

(a)

(b)

Fig. 3.4 **a** Cocrystallization of 1 with 2 yields COF-102 derivatized with allyl groups; **b** subjecting COF-102-allyl to thiol–ene conditions yields COF-102-SPr. Reprinted with permission from Ref. [6]. Copyright 2013, Royal Society of Chemistry

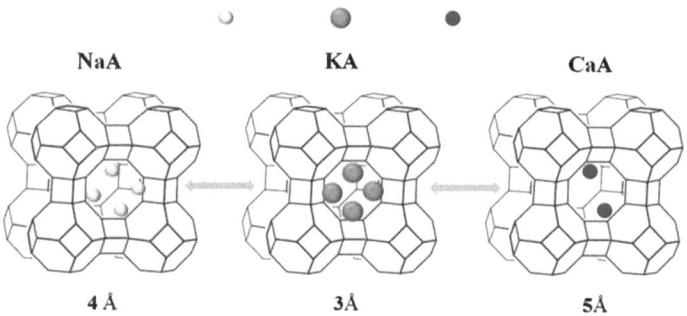

NaA

KA

CaA

4 Å

3 Å

5 Å

Fig. 3.5 The pore size of zeolite A could be tuned by different metal ions

The first POF with charged skeletons was demonstrated by Li et al. [7]. A quaternary phosphonium and tertiary phosphorus functionalized microporous polymer was obtained using tetrakis(4-chlorophenyl)phosphonium bromide as monomer via nickel(0)-catalyzed Yamamoto-type cross-coupling reaction (Fig. 3.6a). The resulting product showed high stability, indicated by treatment in water, base, and acid. The apparent surface areas of the sample could be tuned by exchanging the

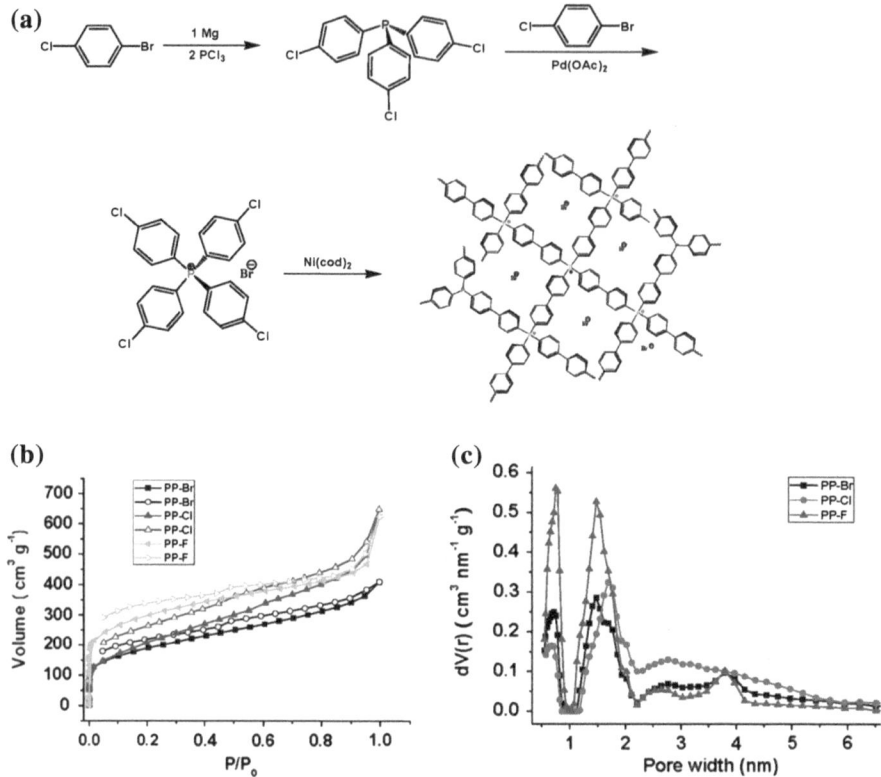

Fig. 3.6 **a** Synthesis of polymer networks; **b** nitrogen adsorption and desorption isotherms measured at 77 K (the adsorption and desorption branches are labeled with solid and open symbols, respectively); **c** pore size distribution of the polymer with different anions. Reprinted with permission from Ref. [7]. Copyright 2012, American Chemical Society

anions from Br$^-$ to F$^-$, from 650 to 980 m^2 g^{-1} (Fig. 3.6b). Interestingly, the ionic microporous polymer displayed high catalytic activity in the reaction between epoxide and CO$_2$ (yield: 98 %, 1 atm).

Recently, our group illustrated the design and synthesis of quaternary pyridinium-type PAFs via condensation of 4-pyridinylboronic acid and cyanuric chloride (Fig. 3.7) [8]. N$_2$ sorption isotherms revealed the microporous character of PAF-50 with narrow pore size distribution centered at 5 Å. Inspired by this the pore size of zeolite A could be tuned by exchanging of cations, combining the ionic skeleton of PAF-50 and its narrow distributed pore size; we attempted to prepare "organic zeolite" by an ion exchange approach [9]. As expected, the series of pyridinium-type PAF-50s with different pore sizes from 3.4 to 7 Å were obtained using anion exchange (Fig. 3.8), which was indicated by the gas sorption isotherms. Their pore sizes are close to the kinetic diameters of light gases, therefore they could capture or sieve gas molecules efficiently by size-exclusive effects. The designed pore sizes may bring benefits to capturing or sieving gas molecules with varied diameters to separate them efficiently by size-exclusive effects. By combining their specific separation

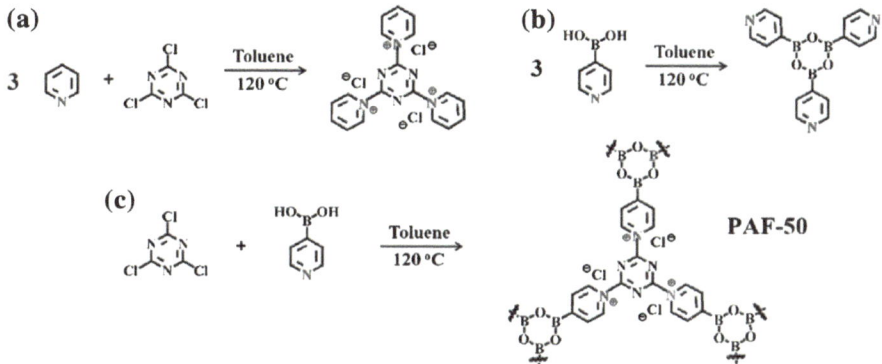

Fig. 3.7 **a**, **b** Condensation reactions used to produce discrete molecules and extended PAFs; **c** scheme of synthesis of PAF-50

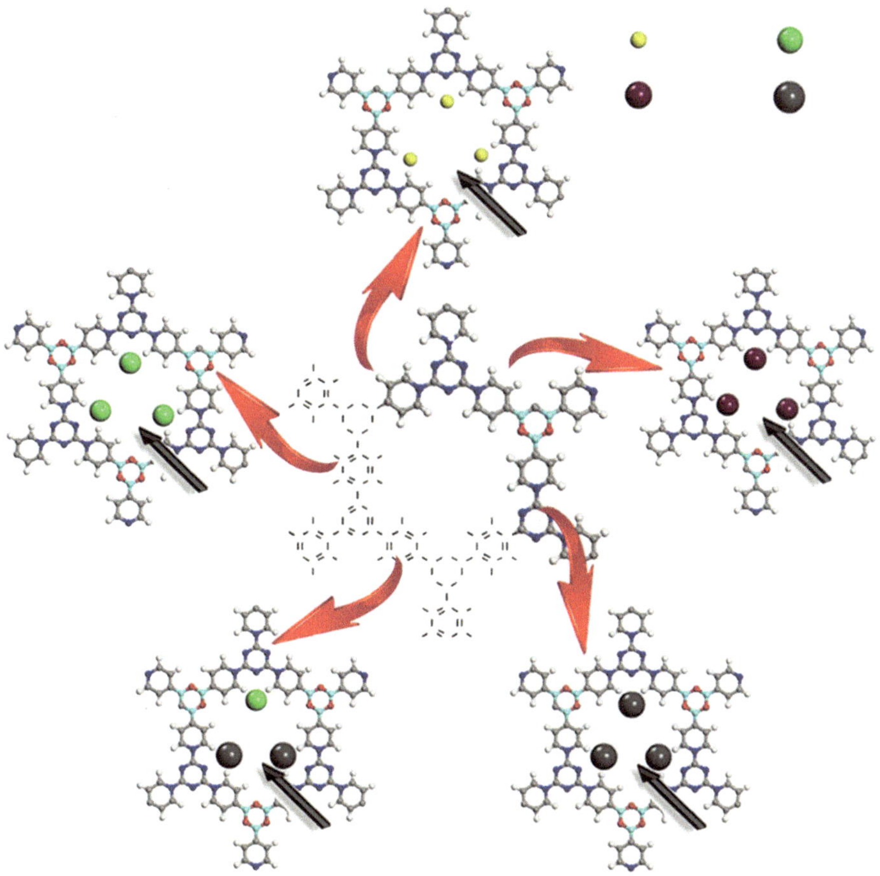

Fig. 3.8 Scheme of preparation of F-PAF-50, Br-PAF-50, 2I-PAF-50, and 3I-PAF-50 from Cl-PAF-50

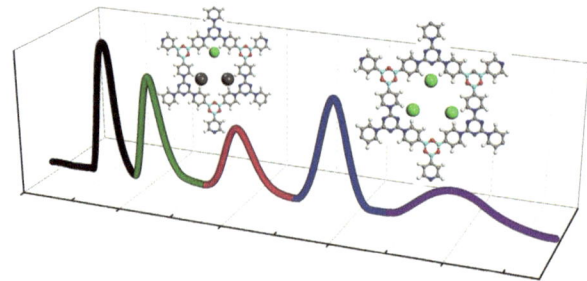

Fig. 3.9 Total separation. GC chromatograms on the Cl-PAF-50 and 2I-PAF-50 connective column for separation of H_2, N_2, O_2, CH_4, and CO_2 mixture

properties, a penta-component (hydrogen, nitrogen, oxygen, carbon dioxide, and methane) gas mixture can be separated completely (Fig. 3.9).

3.2.3 Metallization of Reactive Site

For POFs, one of the most promising applications is in the field of gas storage and separation. To enhance gas uptake, porous materials should have strong interactions with gas molecules. Several strategies in the literature have been developed to enhance interactions between porous materials and gas molecules via the control of pore sizes, incorporation of polar functional groups, or the introduction of open metal sites into the porous networks.

(a)

PAF-18-OH

PAF-18-OLi

(b)

M= Li, Na, K, Mg$_{0.5}$

Fig. 3.10 a Schematic depiction of the synthesis of PAF-18-OH and PAF-18-OLi; **b** Schematic depiction of the synthesis of PAF-26-COOH and PAF-26-COOM

Inspired by open metal sites containing MOFs having high gas adsorption performance, our group attempted to prepare POFs with open metal sites through post-modification approach: (a) We synthesized new porous aromatic frameworks (PAFs) with permanent porosity and functionalized OH groups (PAF-18-OH) (Fig. 3.10a) [10]. The OH groups of PAF-18-OH were further transformed into lithium alkoxy groups, resulting in lithium-loaded PAF, referred to as PAF-18-OLi. As expected, the gas adsorption ability of PAF-18-OLi increased; (b) compared to the OH groups, carboxyl is more reactive. Thereby a new PAF featuring with carboxyl-decorated pores, PAF-26-COOH, has been synthesized successfully (Fig. 3.10b) [11]. The modification of PAF-26 materials with representative light metals was exemplified by Li, Na, K, and Mg via a post-metalation approach. As expected, compared with the parent material PAF-26-COOH, an enhanced adsorption for CO_2 and CH_4 was observed for PAF-26-COOM. The results indicate that introduction of light metal ions would improve the adsorption affinity to gas molecules.

3.3 Direct Approach for Decorating POFs

POFs are composed of organic building units. One vivid characteristic of POFs is that their skeletons contain plentiful phenyl rings. As we know, phenyl ring is a very good precursor of the preparation of many functional group-containing compounds (Fig. 3.11).

One problem of functionalization is that further treatment of phenyl rings is commonly performed under rigorous conditions, such as strong acid, Br_2, high temperature, etc. Thereby, it requires POFs to be stable enough to bear the reaction conditions. In addition, when the reaction occurs, the functional groups will occupy the inner wall of the pores of POF materials. To allow the occupation of functional groups, POFs with high surface area and suitable pore size could be considered as a promising starting material. Our group successfully prepared PAF-1 which possessed exceptionally high BET surface area up to 5,600 $m^2 g^{-1}$ [12]. There was no decrease in the surface area of PAF-1 even when it was treated with hot water for 1 week; TGA shows that decomposing temperature of PAF-1 exceeds 500 °C. Both results indicate that PAF-1 is highly stable. Based on N_2 sorption isotherm, pore size of PAF-1 is 1.4 nm. Combining its high surface area, excellent stability, and suitable pore size, PAF-1 is a promising precursor for synthesis of functional groups containing POFs.

3.3.1 Sulfonic Acid Grafted PAF-1

The first report concerning modifying PAF-1 was illustrated by Zhou et al. PAF-1 is grafted with chlorosulfonic acid, and the resulting product is sulfonic acid contained PAF-1 (Fig. 3.12) and then decorated with lithium salt (PAF-1-SO_3Li) [13]. Alternatively, concentrated sulfonic acid could also achieve the sulfonation reaction [14]. PAF-1-SO_3NH_4 could be easily obtained by simply mixing PAF-1-SO_3H and ammonia hydroxide.

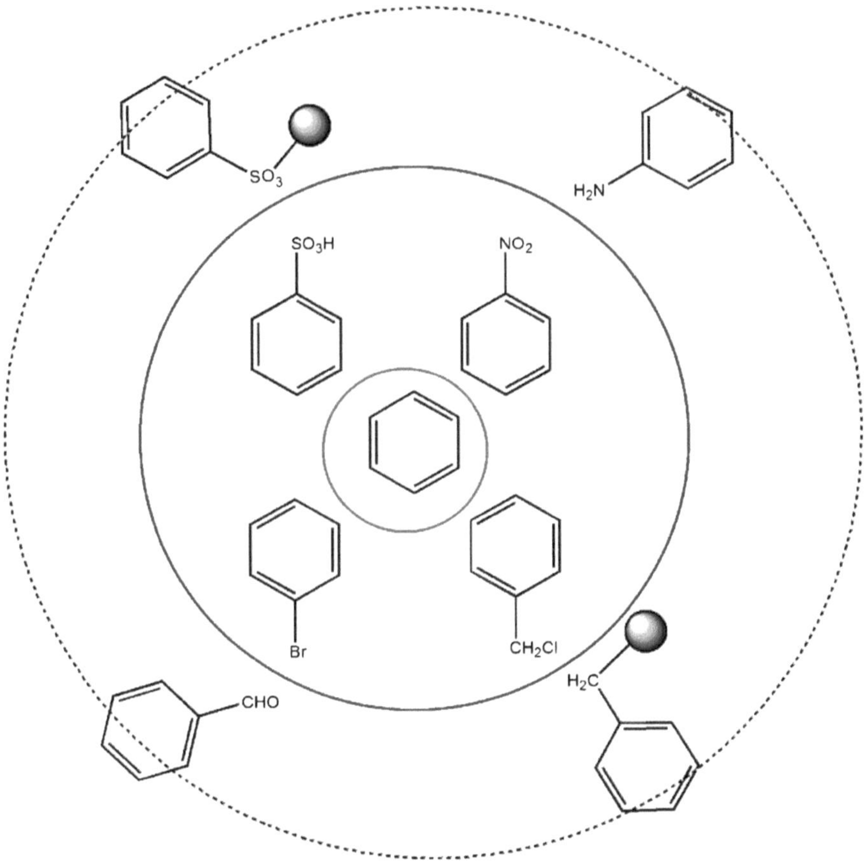

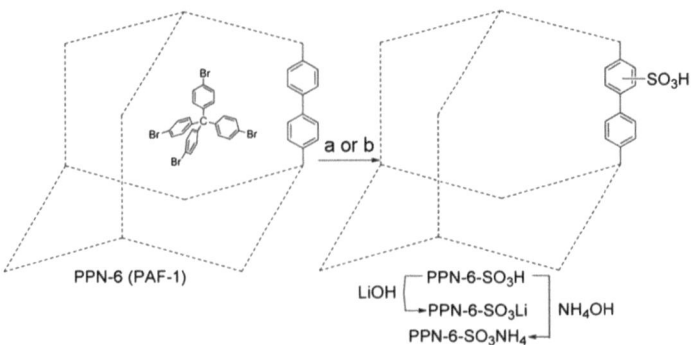

Fig. 3.11 Typical strategies for functionalizing phenyl rings. When certain reactive groups are decorated, further modification could be performed

Fig. 3.12 Synthetic route for sulfonate functionalized PPNs. Chlorosulfonic acid or concentrated sulfonic acid could be used for the sulfonation reaction. Reprinted with permission from Ref. [14]. Copyright 2013, Royal Society of Chemistry

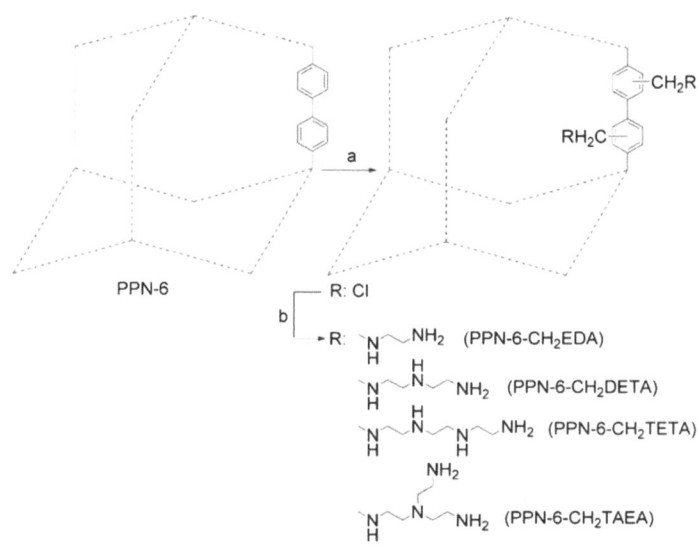

Fig. 3.13 Synthetic route to polyamine-tethered PPNs. Reproduced from Ref. [15] by permission of John Wiley & Sons Ltd.

3.3.2 Poly-amine Substitution of PAF-1

Additionally, the same group prepared a family of poly-amine substituted PAF-1 to obtain adsorbents with high CO_2 capacity and selectivity [15]. First, PAF-1 was performed chloromethylation using $CH_3COOH/HCl/H_3PO_4/HCHO$ system. The resulting PPN-6-CH_2Cl could further react with different amine compounds, and formed poly-amine grafted PAF-1 (Fig. 3.13). In comparison with PAF-1, calculated BET surface areas of various poly-amine grafted PAF-1 materials dramatically decrease with values ranging between 555 and 1,014 $m^2\ g^{-1}$. As expected, the poly-amine grafted PAF-1 materials possess significantly enhanced CO_2 uptake and decreased N_2 uptake capacity at low pressures and 295 K. Among those poly-amine grafted PAF-1 materials, PAF-1-CH_2DETA exhibits ultrahigh binding affinity and ever largest selectivity for CO_2 compared to any reported porous material.

3.3.3 Dual Functionalization of PAF-1

The capability of enzymes could allow a cascade reaction, performing a consecutive series of reactions. The development of bi/multifunctional solid catalysts for heterogeneous cascade catalysis is required in chemical industries.

Ma et al. demonstrated the functionalization of PAF-1 (Fig. 3.14) [16]. To achieve the cascade reaction, two antagonistic sites of strong acid and strong base were readily grafted to the phenyl rings via stepwise post-synthetic modification

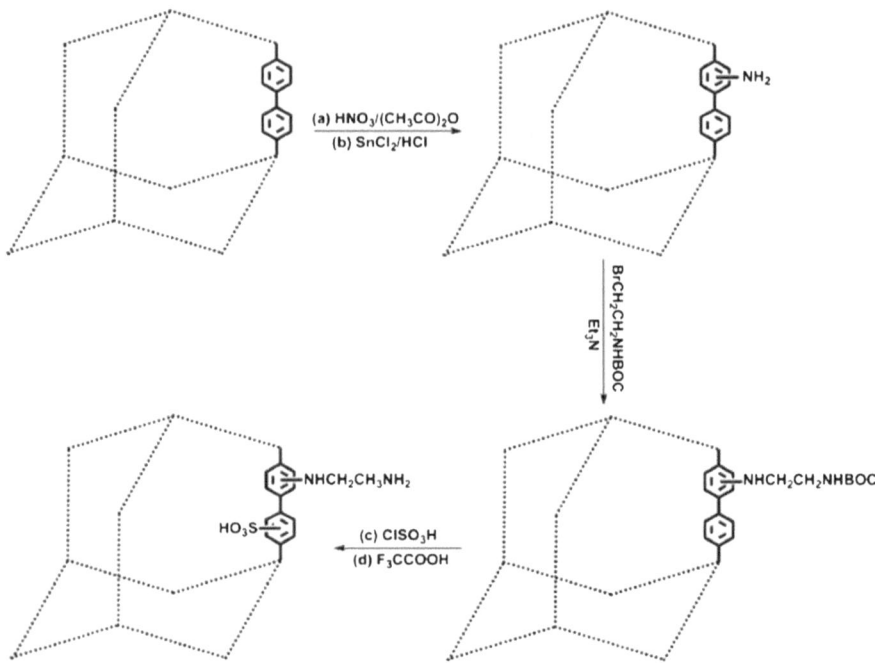

Fig. 3.14 Illustration of PAF-1 to graft the strong acid and base to afford PAF-1-NHCH₂CH₂NH₂-SO₃H. Reprinted with permission from Ref. [16]. Copyright 2014, Royal Society of Chemistry

using selected organic reactions. As expected, the resulting bifunctionalized PAF-1 showed excellent performances in catalyzing a series of cascade reactions.

In summary, the structure and property of POFs could be modified using the post-synthetic approach. If the POFs have reactive groups, the modification is relatively facile. Additionally, as indicated by the modification of PAF-1, the high surface area and stability of POFs provide an indirect approach to achieve the target.

References

1. Xiang Z, Cao D (2013) Porous covalent–organic materials: synthesis, clean energy application and design. J Mater Chem A 1:2691–2718
2. Patton GC (2004) Development and applications of click chemistry. November 8, 2004. Online http://www.scs.uiuc.edu
3. Nagai A, Guo Z, Feng X et al (2011) Pore surface engineering in covalent organic frameworks. Nat Commn 2:536–543
4. Xu H, Chen X, Gao J et al (2014) Catalytic covalent organic frameworks via pore surface engineering. Chem Commun 50:1292–1294
5. Cote A, Benin A, Ockwig N et al (2005) Porous, crystalline, covalent organic frameworks. Science 310:1166–1170

6. Bunck D, Dichtel W (2013) Postsynthetic functionalization of 3D covalent organic frameworks. Chem Commun 49:2457–2459

7. Zhang Q, Zhang S, Li S (2012) Novel functional organic network containing quaternary phosphonium and tertiary phosphorus. Macromolecules 45:2981–2988

8. Yuan Y, Sun F, Zhang F et al (2013) Targeted synthesis of porous aromatic frameworks and their composites for versatile, facile, efficacious, and durable antibacterial polymer coatings. Adv Mater 25:6619–6624

9. Yuan Y, Sun F, Li L et al (2014) Porous aromatic frameworks with anion-templated pore apertures serving as polymeric sieves. Nat Commun 5:4260–4267

10. Ma H, Ren H, Zou X et al (2013) Novel lithium-loaded porous aromatic framework for efficient CO_2 and H_2 uptake. J Mater Chem A 1:752–758

11. Ma H, Ren H, Zou X et al (2014) Post-metalation of porous aromatic frameworks for highly efficient carbon capture from $CO_2 + N_2$ and $CH_4 + N_2$ mixtures. Polym Chem 5:144–152

12. Ben T, Ren H, Ma S et al (2009) Targeted synthesis of a porous aromatic framework with high stability and exceptionally high surface area. Angew Chem Int Ed 48:9457–9460

13. Lu W, Yuan D, Sculley J et al (2011) Sulfonate-grafted porous polymer networks for preferential CO_2 adsorption at low pressure. J Am Chem Soc 133:18126–18129

14. Lu W, Verdegaal W, Yu J et al (2013) Building multiple adsorption sites in porous polymer networks for carbon capture applications. Energ Environ Sci 6:3559–3564

15. Lu W, Sculley J, Yuan D et al (2012) Polyamine-tethered porous polymer networks for carbon dioxide capture from flue gas. Angew Chem Int Ed 51:7480–7484

16. Zhang Y, Li B, Ma S (2014) Dual functionalization of porous aromatic framework as a new platform for heterogeneous cascade catalysis. Chem Commun 50:8507–8510

Chapter 4
Gas Sorption Using Porous Organic Frameworks

Abstract POFs possess high stability, high specific surface area, and adjustable structural framework. Therefore, POFs could be considered as a desired media for gas storage. Specific to the energy and environment fields, it has been documented that POFs are employed in the fields of clean fuels storage, greenhouse gases capture, and pollution elimination. In this chapter, we introduce the application of POFs for hydrogen, methane, carbon dioxide, and small hydrocarbon storage. The factors influencing gas adsorption capacities are discussed and summarized, including the surface area, pore size, pore surface polarity, etc.

Keywords Hydrogen storage · Methane storage · Carbon dioxide capture · Small hydrocarbon storage · Influencing factors

4.1 Introduction

POF materials have become a hot issue for their important roles in storage and separation applications [1–3]. Specific to energy and environment fields, POFs could be applied for storage of clean fuels, capture of greenhouse gases, and elimination of pollutions. In this chapter, we introduce the application of POFs for hydrogen, methane, carbon dioxide, and small hydrocarbon storage. Through analyzing the data of gas capacities, the influencing factors are discussed and summarized.

4.2 Hydrogen Storage

Due to its clean combustion product and high chemical energy density, hydrogen as an energy source is proposed as an alternative to traditional fossil fuels [4]. However, large-scale storage and transport of hydrogen in a safe and compact

© The Author(s) 2015   57
G. Zhu and H. Ren, *Porous Organic Frameworks*, SpringerBriefs in Green
Chemistry for Sustainability, DOI 10.1007/978-3-662-45456-5_4

way is still highly desired. One strategy is chemisorption: hydrogen is stored in the form of metal hydrides. This strategy has been widely studied, but there are still some problems due to their high storage and release temperature. Another strategy is physisorption. Porous materials with high surface area are widely studied because of their rapid uptake and release of hydrogen. To date, there are thousands of articles on H_2 storage. Porous materials, such as activated carbon [5], zeolites [6], and metal–organic frameworks (MOFs) [7] are employed for this application. As a new family of porous materials, POFs composed of light elements (C, H, O, N, B, etc.) via covalent bonds [8, 9], have attracted a great deal of attention thanks to their high surface areas, high stability, and controllable skeletons.

Along with the increasing utilization of the world's fast diminishing fossil fuels, replacing them by using clean energy is extremely urgent. Materials that have sufficiently large H_2 storage capacity are highly desired for the practical application of H_2 utilization. The 2015 H_2 storage target set by the U.S. Department of Energy (DOE) for hydrogen fuel cell vehicles is 5.5 wt% in gravimetric capacity, at an operating temperature of 233–333 K and under a maximum delivery pressure of 100 bar [10].

As listed in Table 4.1, the H_2 storage abilities of typical POFs are compared. The measurement is usually performed at low temperature (77 K), and under both low and high pressure. The data present some very useful information for our investigations: (1) H_2 storage capacity is influenced by the temperature, at room temperature the H_2 storage is very low; (2) under low pressure, the H_2 storage capacities of most POFs are lower than 2 wt%, and they are independent of the surface areas of POFs; (3) under high pressure, the H_2 storage capacities increase with the increasing surface areas of POFs; (4) the modification of the skeleton of POFs could tune the H_2 storage ability. Later, we introduce the strategies for enhancing the H_2 storage ability.

4.2.1 Small Pore Effect

For hydrogen, its interaction with porous materials is very weak. If the size of guest molecules could match the host materials, the interaction between guest and host would be enhanced. One distinguished example is PAF-1 (BET surface area 5,640 m^2 g^{-1}), the H_2 capacity of which is lower than that of other POFs [29]. The pore size of PAF-1 reaches 1.3 nm, and it is too large for H_2 located inside the pores under low pressure. The pore size of POFs affecting the H_2 storage ability was proved by the work of Han et al. [20]. Microporous polycarbazole was synthesized via straightforward carbazole-based oxidative coupling polymerization (Fig. 4.1a). CPOP-1 has high BET surface area with value of 2,220 m^2 g^{-1} and narrow pore size distribution centered at about 0.62 nm (Fig. 4.1b). The hydrogen storage is 2.80 wt% (1.0 bar and 77 K). The micropores dominated CPOP-1 is an excellent media for H_2 storage.

Table 4.1 Hydrogen storage in POFs

Material	S_{BET} (m²/g)	V (cm³/g)	P (bar)	T (K)	wt%	References
COFs						
COF-1	750	0.3	1	77	1.46	[11]
COF-1	628	0.36	1	77	1.28	[12]
			100	298	0.26	
COF-5	1,670	1.07	1	77	3.54	[11]
COF-6	750	0.32	1	77	2.23	[11]
COF-8	1,350	0.69	1	77	3.46	[11]
COF-10	1,760	1.44	1	77	3.88	[11]
COF-102	3,620	1.55	1	77	7.16	[11]
			42	77	7.24	[11]
COF-103	3,530	1.54	1	77	6.98	[11]
			42		7.05	[11]
COF-18 Å	1,263	0.69	1	77	1.55	[13]
COF-16 Å	753	0.39	1	77	1.4	[13]
COF-14 Å	805	0.41	1	77	1.23	[13]
COF-11 Å	105	0.052	1	77	1.22	[13]
ILCOF-1	2,723	1.21	1	77	1.3	[14]
			40	77	4.7	
TDCOF-5	2,497	1.3	1	77	1.6	[15]
CTV-COF-1	1,245	0.93	1	77	1.3	[16]
CTV-COF-2	1,170	1.07	1	77	0.75	[16]
PIMs						
Phthalocyanine	895					[17]
Porph-PIM	980		1	77	1.2	[17]
			10	77	1.95	
HATN-PIM	850		1	77	1.12	[17]
			10	77	1.56	
PIM-1	850	0.78	1	77	0.95	[17]
			10	77	1.45	
CTC-PIM	830		1	77	1.35	[17]
			10	77	1.7	
Trip-PIM	1,064		1	77	1.63	[18]
			10	77	2.71	
Others						
OFP-3	1,159		1	77	1.56	[19]
			10	77	3.94	
CPOP-1	2,220		1	77	2.8	[20]
CPOP-2	510		1	77	1.29	[21]
CPOP-3	630		1	77	0.91	[21]
CPOP-4	660		1	77	0.93	[21]
CPOP-5	1,050		1	77	1.26	[21]

(continued)

Table 4.1 (continued)

Material	S_{BET} (m²/g)	V (cm³/g)	P (bar)	T (K)	wt%	References
CPOP-6	980		1	77	1.23	[21]
CPOP-7	1,430		1	77	1.51	[21]
EOF-1	780		1	77	0.94	[22]
EOF-2	1,046		1	77	1.21	[22]
EOF-3	445	0.27	1	77	0.55	[23]
EOF-4	423	0.27	1	77	0.56	[23]
EOF-5	261	0.19	1	77	0.26	[23]
EOF-6	1,380	0.9	1	77	1.29	[23]
EOF-7	1,083	0.69	1	77	1.14	[23]
EOF-8	540	0.32	1	77	0.72	[23]
EOF-9	602	0.38	1	77	0.86	[23]
CMPs						
CMP-0	1,018	0.56	1.13	77	1.4	[24]
			1.13	293	0.006	[25]
Pd-CMP-0	604	0.36	1.13	293	0.069	[25]
CMP-1	834	0.47	1.13	77	1.14	[24]
CMP-2	634	0.53	1.13	77	0.92	[24]
Li-CMP	795	1.61	1	77	6.1	[26]
			1	273	0.1	
BLP-1(Cl)	1,364	0.75	1	77	1.1	[27]
BLP-1(Br)	503	0.33	1	77	0.68	
BLP-2(Cl)	1,174	0.65	1	77	1.3	[27]
BLP-2(Br)	520	0.57	1	77	1.98	
BLP-2(H)	1,178	0.59	1	77	1.4	[28]
			15	77	2.48	
PAF-1	5,600	3.05	1	77	1.66	[29]
			48	77	7	[30]
Li@PAF-1	479		1.2	77	2.7	[31]
PAF-3	2,932		1	77	2.07	[29]
			60	77	5.5	
PAF-4	2,246		1	77	1.5	[29]
			60	77	4.2	
PPN-4	6,461	3.04	55	77	9.1	[32]
PAF-18-OH	1,121	0.82	1	77	1.35	[33]
PAF-18-OLi	981	0.57	1	77	1.65	[33]
POP-1	1,031	0.64	60	77	2.78	[34]
			60	87	2.31	
POP-2	1,012	0.71	60	77	2.71	
			60	87	2.14	
POP-3	1,246	0.73	60	77	3.07	
			60	87	2.51	

(continued)

Table 4.1 (continued)

Material	S_{BET} (m²/g)	V (cm³/g)	P (bar)	T (K)	wt%	References
POP-4	1,033	0.73	60	77	2.35	
			60	87	1.75	
BILP-1	1,172		1	77	1.9	[35]
BILP-2	708	0.49	1	77	1.3	
BILP-3	1,306		1	77	2.1	
BILP-4	1,135	0.65	1	77	2.3	
BILP-5	599	0.36	1	77	1.4	
BILP-6	1,261		1	77	2.2	
BILP-7	1,122	0.74	1	77	1.8	
HCPs						
Davankov resins	1,466		1	77	1.28	[36]
			10	77	2.75	
			15	77	3.04	
	2,090		1.2	77	1.55	[36]
pDCX	1,370		1.13	77	1.69	[37]
BCMBP/DCX	1,904		1.13	77	1.61	[37]
Precursor polyanilines	632	0.94	1.2	77	0.96	[38]
			3.0	77	2.2	[38]
	480	0.55	1.2	77	0.82	[38]
	54	0.13	1.2	77	0.38	[39]
	316	0.25	1.2	77	0.85	[39]
Aniline	249	0.13	1.2	77	0.97	[39]

4.2.2 Simulating the H₂ Uptake

By employing grand canonical Monte Carlo (GCMC) simulations, Goddard reported the H_2 uptake for COF-102, COF-103, and COF-202 at 298 K. In addition, they also proposed a new strategy to obtain COFs with higher interaction with H_2, which is metalating the COF with alkali metals [40]. Excitingly, the resulting materials show exceptionally high H_2 uptakes at 298 K, with value of COF-102-Li (5.16 wt%), COF-103-Li (4.75 wt%), COF-102-Na (4.75 wt%), and COF-103-Na (4.72 wt%), which all exceed the DOE target. Other strategies to increase the interaction of COFs with molecular hydrogen have been proposed recently [41]. The new COF-301-PdCl$_2$ is predicted to reach 60 g total H_2/L at 100 bar (Fig. 4.2), which is 1.5 times of the DOE 2015 target of 40 g/L and close to the ultimate (2050) target of 70 g/L [42].

Recently, Cao et al. theoretically predicted two new tetrahedral node diamondyne frameworks by replacing the carbon nodes of diamondyne and diamond with the acetylenic linkage-formed tetrahedron node, marked as TND-1 and TND-2 (Fig. 4.3) [43]. The specific surface area of TND-1 and TND-2 is 6,250 and

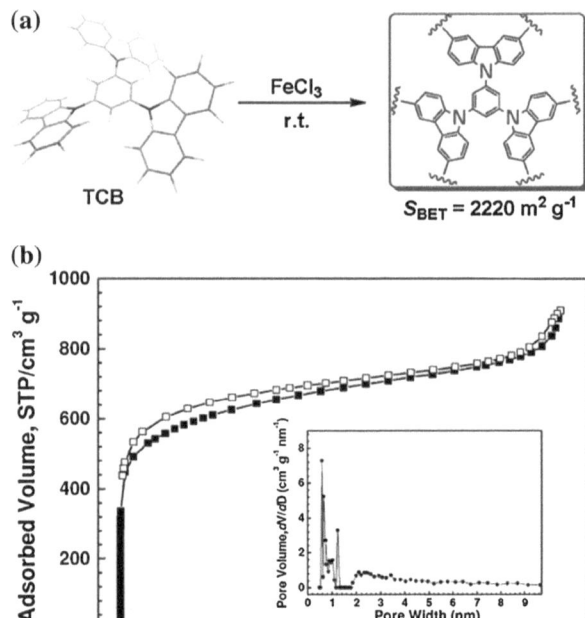

Fig. 4.1 a Synthetic route to microporous polycarbazole CPOP-1; **b** nitrogen adsorption–desorption isotherms and the pore size distribution calculated by NLDFT (*inset*) of CPOP-1. Reprinted with permission from Ref. [20]. Copyright 2012, American Chemical Society

$2{,}992$ m^2 g^{-1}, respectively. Excitingly, at 298 K and 100 bar, the H$_2$ adsorption capacity of TND-1 and TND-2 is 27 and 13 mg g^{-1}. This prediction of POF structures would direct the design of POF.

4.2.3 Lithium Modified POFs

The first study concerning Li-doped POFs was reported by Deng et al. [26]. CMP is used as starting material, which is synthesized by PdII/CuI-catalyzed homocoupling polymerization. Naphthalene anion radical salt was utilized for the doping experiments. The resulting Li-CMP with 0.5 wt% Li exhibits a significantly high adsorption of maximum 6.1 wt% at 1 bar, which is nearly four times that of the CMP (Fig. 4.4). This is close to POFs with high surface area and measured at high pressure.

In 2009, our group reported the design and synthesis of PAF-1, which showed exceptionally high surface areas. At 77 K and 48 bar, its H$_2$ adsorption capacity reached 7.0 wt% [30]. A series of PAFs that shared similar topology were also obtained subsequently. At 60 bar and 77 K, H$_2$ uptake of PAF-3 and PAF-4 reached 5.5 wt% and 4.2 wt%, respectively. Matthew et al. presented a route for lithiation of PAF-1 (Li@PAF-1) (Fig. 4.5), resulting in an activated pore surface. H$_2$ adsorption isotherms collected at 77 K and 1.2 bar exhibited a storage capacity of 2.7 wt% for 5 %-Li@PAF-1 [31].

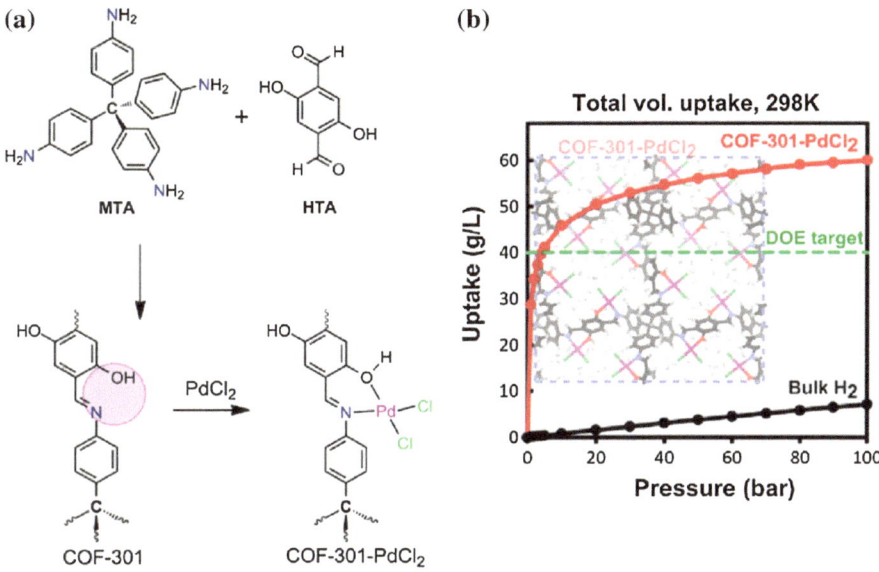

Fig. 4.2 a Synthetic route to COF-301-PdCl$_2$; **b** total volumetric uptake H$_2$ isotherms obtained at 298 K for COF-301-PdCl$_2$. Reprinted with permission from Ref. [42]. Copyright 2012, American Chemical Society

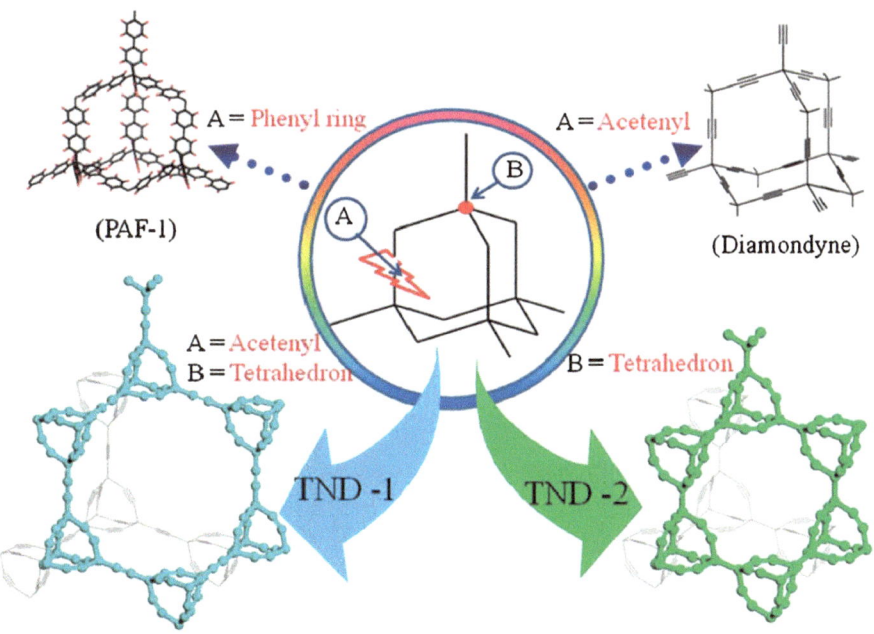

Fig. 4.3 The diamond-like frameworks for PAF-1, diamondyne, TND-1 and TND-2, respectively. Reprinted with permission from Ref. [43]. Copyright 2014, Royal Society of Chemistry

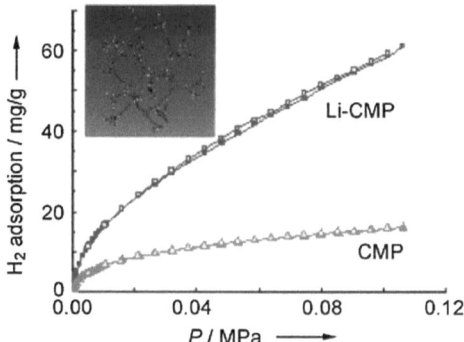

Fig. 4.4 Comparison of Li-CMP and CMP in H_2 uptake. Reprinted with permission from Ref. [26]. Copyright 2010, Wiley-VCH

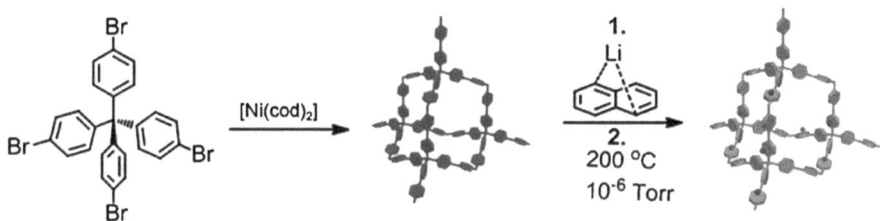

Fig. 4.5 Synthesis of Li@PAF-1 materials. Reprinted with permission from Ref. [31]. Copyright 2012, Wiley-VCH

Recently, we synthesized a new PAF material with permanent porosity and functionalized OH groups, PAF-18-OH [33]. The OH groups of PAF-18-OH were further transformed into lithium alkoxy groups, referred to as PAF-18-OLi. The H_2 uptake of PAF-18-OH is 1.35 wt% at 77 K, and 0.95 wt% at 87 K at 1 bar, respectively (Fig. 4.6c). Compared with PAF-18-OH, the surface area of PAF-18-OLi decreases (Fig. 4.6a), but its H_2 uptake increases. PAF-18-OLi with Li content of 4.2 wt% shows a maximum uptake of 1.65 wt% at 77 K, and 1.17 wt% at 87 K at 1 bar, respectively.

Although the H_2 uptakes of some POFs reach $\sim$7 wt% at 77 K (COF-102, COF-103 [11], PAF-1, and PPN-4 [32], the use of these materials for H_2 storage would still appear to be far away due to the cryogenic temperatures that are required. The methods suggested by theoretical method to enhance the hydrogen adsorption enthalpy, such as metal intercalation, are yet to be borne out in practice.

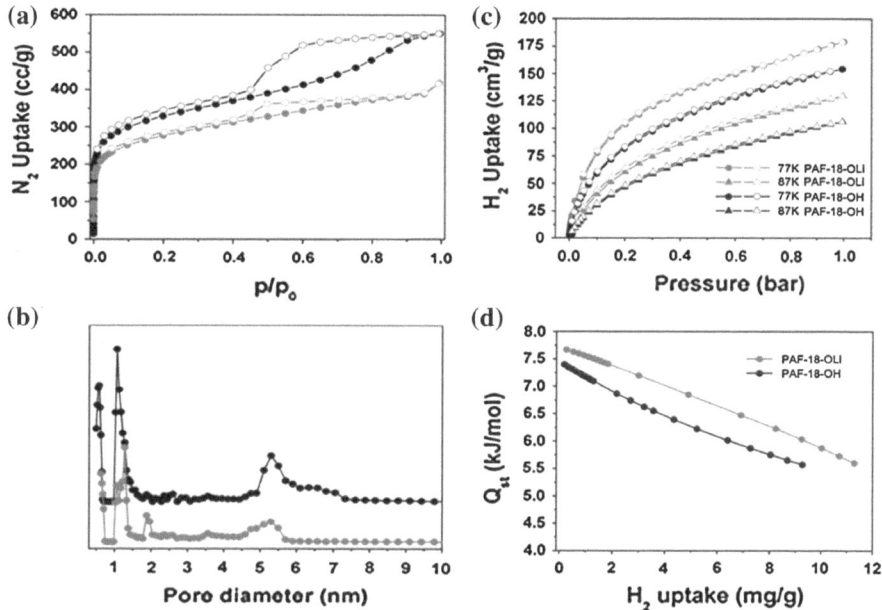

Fig. 4.6 **a** N_2 adsorption and desorption isotherms of PAF-18-OH (*black symbols*) and PAF-18-OLi (*red symbols*) at 77 K; **b** the pore size distribution of PAF-18-OH (*black circles*) and PAF-18-OLi (*red circles*); **c** H_2 adsorption and desorption isotherms of PAF-18-OH and PAF-18-OLi at 77 and 87 K; **d** plots of H_2 adsorption Qst of PAF-18-OH and PAF-18-OLi versus the H_2 uptake

4.3 Methane Storage Using POFs

Besides the renewable energy source H_2, CH_4 also seems to be a rising star in clean energy, with the power density of 15.5 kW h kg^{-1} and a worldwide storage. Currently, the main problem with CH_4 application lies in the efficient storage and transport of this highly flammable gas. Concerns over the stability and cost associated with the application of MOFs in methane adsorption have led to the evaluation of POFs as alternative methane sorbents. Their tolerance of water and metal-free skeletons make POFs attractive options in the field of methane storage. Furthermore, many POFs exhibit exceptionally high surface areas and low framework density, which make them ideal in the gravimetric storage of methane. The investigations on methane adsorption in nanoporous organic polymers are much less than that of widely studied hydrogen sorption. To satisfy the demand on CH_4 transport, the materials should have high CH_4 adsorption capacities.

4.3.1 Temperature and Pressure Effect

Tables 4.2 and 4.3 list the methane uptake of some representative POFs under low and high pressure, respectively. At low pressure, CH_4 uptakes of POFs are relatively low with values below 1.7 mmol g^{-1}. The CH_4 uptakes of POFs are independent of their surface area. In addition, from the data collected at 273 K and 298 K, along with the increasing temperature, the CH_4 uptake of POFs greatly decreases, indicating that temperature is an important factor influencing CH_4 adsorption ability.

When measurement is performed under high pressure, the CH_4 uptakes of POFs reach a promising level. Yaghi et al. reported a series of COFs for the CH_4 storage [11]. Among these POFs, PPN-4 with 6,461 m^2 g^{-1} BET surface area had the highest CH_4 uptake with a value of 17.1 mmol g^{-1}, at 55 bar and 295 K.

Table 4.2 Methane storage in POFs under low pressure

Material	S_{BET} (m^2/g)	P (bar)	T (K)	CH_4 uptake (mmol/g)	References
PAF-1	5,600	1	273	1.12	[29]
Li@PAF-1	479	1.22	273	1.3	[31]
PAF-3	2,932	1	273	1.68	[29]
PAF-4	2,246	1	273	1.12	[29]
PPF-1	1,740	1	273	1.52	[44]
PPF-2	1,470	1	273	1.44	[44]
PPF-3	419	1	273	0.63	[44]
PPF-4	726	1	273	0.83	[44]
PAF-41	1,119	1	273	1.04	[45]
			298	0.68	[45]
PAF-42	640	1	273	0.68	[45]
		1	298	0.34	[45]
PAF-43	515	1	273	0.60	[45]
		1	298	0.28	[45]
PAF-44	532	1	273	0.66	[45]
		1	298	0.41	[45]
BILP-2	708	1	273	0.88	[35]
		1	298	0.56	[35]
BILP-4	1,135	1	273	1.63	[35]
		1	298	1.13	[35]
BILP-5	599	1	273	0.94	[35]
		1	298	0.63	[35]
BILP-7	1,122	1	273	1.63	[35]
		1	298	1.13	[35]
BILP-3	1,306	1	273	1.50	[46]
		1	298	1.06	[46]
BILP-6	1,261	1	273	1.69	[46]
		1	298	1.19	[46]

Table 4.3 Methane storage in POFs under high pressure

Material	SA_{BET} (m^2/g)	P (bar)	T (K)	CH_4 uptake (mmol/g)	References
COFs					
COF-1	750	35	298	2.5	[11]
COF-5	1,670	35	298	5.6	[11]
COF-6	750	35	298	4.1	[11]
COF-8	1,350	35	298	5.4	[11]
COF-10	1,760	35	298	5.0	[11]
COF-102	3,620	35	298	11.7	[11]
COF-103	3,530	35	298	10.9	[11]
HCPs					
1	1,904	15	298	4.4	[47]
		20	298	5.2	
2	1,307	15	298	3.9	[47]
		20	298	4.5	
3	963	15	298	3.2	[47]
		20	298	3.5	
		36	298	4.3	
4	1,366	15	298	4.3	[47]
		20	298	4.6	
		36	298	5.5	
PPN-1	1,249	70	295	6.25	[48]
PPN-2	1,764	70	295	9.4	[48]
PPN-3	4,221	70	295	12.2	[48]
PPN-4	6,461	55	295	17.1	[32]
ALP-1		70	298	9.0	[49]
ALP-2		70	298	9.6	[49]
ALP-3		70	298	9.9	[49]
ALP-4		70	298	7.2	[49]

4.3.2 Simulating CH₄ Uptake

Yaghi et al. reported several COFs as exceptional methane storage materials [11]. The best COF in terms of total volume of adsorbed CH_4 per unit volume of COF absorbent is COF-1, which can store 195 v/v at 298 K and 30 bar, exceeding the DOE target for CH_4 storage of 180 v/v at 298 K and 35 bar. The best COFs for methane adsorption are COF-102 and COF-103 with values of 230 and 234 v/v (from 5 to 100 bar), respectively, making these promising materials for practical methane storage [50]. TND-1 and TND-2 were also theoretically predicted by Cao et al. to have CH_4 uptake abilities [43]. It is noted that, at 35 bar and 298 K, the CH_4 uptake of TND-1 is 350 mg g^{-1}, which is higher than 238 mg g^{-1} of PAF-302, 193 mg g^{-1} of IRMOF-14, and even 253 mg g^{-1} of PCN-14 (290 K).

4.3.3 Metal Modified POFs

The CH_4 uptake ability is related to the surface area of POFs, measurement pressure, and temperature. One research has attracted our attention, at 1 bar and 273 K, the CH_4 uptake of PAF-1 is 1.12 mmol g^{-1} [29]. Matthew reported that 5 %_Li@PAF-1 exhibited higher CH_4 storage capacities with value of 1.30 mmol g^{-1} (273 K, 1.22 bar) [31]. However, the surface area of 5 %_Li@PAF-1 is only 479 m^2 g^{-1}, which is extremely lower than 5,600 m^2 g^{-1} of PAF-1. Thus, it provides very useful information for us to design and synthesize effective adsorbents to satisfy the CH_4 storage requirement.

Recently, we have successfully designed and synthesized a carboxyl-functionalized PAF material, PAF-26-COOH [51]. Post-metalation of PAF-26-COOH yields a series of PAF-26-COOM derivatives (M=Li, Na, K, Mg). The porosity and pore size are tuned and achieved via this post-metalation method. N_2 adsorption measurements indicate the surface areas of PAF-26-COOM derivatives decrease compared with original PAF-26-COOH (Fig. 4.7a). The dependence of uptakes per effective V_{total} versus pressure for the PAF-26 series toward CH_4 is also calculated. It can be found that the CH_4 uptakes per effective V_{total} of PAF-26-COOM increased, with values from 34 mg cm^{-3} (PAF-26-COOH) to 46 mg cm^{-3} (PAF-26-COOLi, 35 % increased), 54 mg cm^{-3} (PAF-26-COOMg 35 % increased), 56 mg cm^{-3} (PAF-26-COONa, 65 % increased) and 60 mg cm^{-3} (PAF-26-COOK, 76 % increased) (Fig. 4.7b). Based on the above results, we can conclude that introduction of metal active centers would greatly promote gas adsorption capacity.

Compared with H_2 storage, CH_4 storage has promising feasibility to achieve the DOE target because CH_4 storage could be performed at room temperature. For practical application, the POFs should satisfy some basic demands such as high surface area, inexpensive synthesis, mass production under mild condition, etc. In addition, the design and synthesis of POFs could be conducted by theoretical simulation.

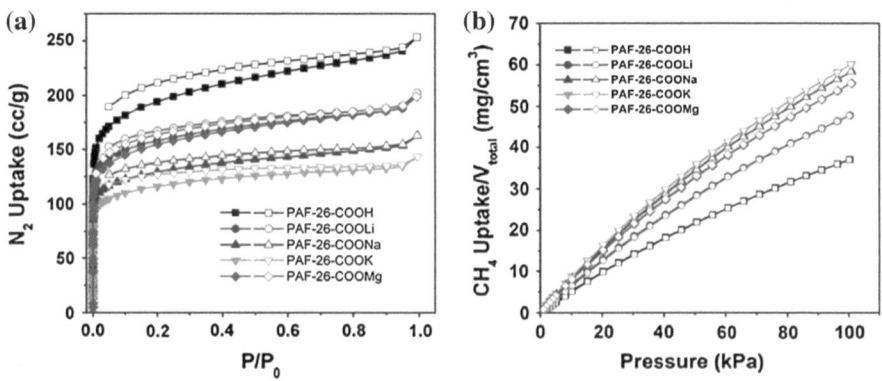

Fig. 4.7 **a** N_2 adsorption–desorption isotherms measured at 77 K for PAF-26-COOH, PAF-26-COOLi, PAF-26-COONa, PAF-26-COOK and PAF-26-COOMg; **b** CH_4 uptakes per effective V_{total} for PAF-26-COOH, PAF-26-COOLi, PAF-26-COONa, PAF-26-COOK and PAF-26-COOMg samples at 273 K and 101 kPa

4.4 Capture and Separation of Carbon Dioxide with POFs

Accelerated by the worldwide economic growth and industrial development, the demand for fossil fuels (such as coal, oil, and natural gas, etc.) is projected to continue to increase in the future. One of the most severe environment concerns of our age is the escalating level of atmospheric greenhouse gases (mainly CO_2) because of the increasing carbon dioxide emission. Subsequent environmental degeneration and adverse climate change sharply affect our civilization today. Flue gas emissions of power plants are responsible for roughly 30 % of total CO_2 emissions. Nitrogen is a main component (>70 %) of flue gas, whereas the major impurity is CO_2 (10–15 %); separating CO_2 from N_2 is highly demanded [52, 53]. Thus, it requires the development of new technologies for capture and sequestration of CO_2 (CCS). The proposed alternative strategies for CCS have been developed so far: (1) chemisorption using aqueous solutions of amines [54]; (2) physical adsorption by porous materials. In the library of solid adsorbents [52, 53], zeolites, carbons, mesoporous silica-supported amines, and metal organic frameworks, have exhibited good performances for practical CCS implementation.

As a steadily developing porous material, porous organic frameworks (POFs) have attracted much attention thanks to their distinctive advantages such as high thermal and chemical stability, tunable pore surface, high surface areas, etc. Therefore, POFs is an excellent candidate for CO_2 capture. Typical postcombusion flue gas mainly contains N_2 (73–77 %), and CO_2 concentration is relatively high (15–16 %) compared to other minor components, such as H_2O, O_2, CO, NOx, and SOx. Facing such complicated gas composition of flue gas, an ideal adsorbent for capturing CO_2 from post-combustion flue gas would display corresponding features: (1) high CO_2 loading capacity; (2) high selectivity for CO_2 over the other flue gas components; (3) long-term stability under rigorous conditions, especially water stability; (4) minimal energy penalty for regeneration; (5) cost for synthesis of POFs, etc.

Table 4.4 lists the CO_2 uptake of some representative POFs under low and high pressure, respectively. Under high pressure, the CO_2 uptake capacity is related to the surface area of POF materials. But under low pressure, the CO_2 uptake capacity is independent of the surface area of POF materials, and the reason is complicated. We introduce some typical strategies to enhance the CO_2 adsorption ability.

4.4.1 Porosity of POFs Controlled by Original Building Units

In 2011, Cooper et al. reported that a series of conjugated microporous polymer (CMP) networks were constructed using the Sonogashira–Hagihara cross-coupling reaction of 1,3,5-triethynylbenzene with different reactive –Br contained compounds [55]. Thus, a range of chemical functional groups including carboxylic acids, amines, hydroxyl groups, and methyl groups have been successfully

Table 4.4   Carbon dioxide storage in POFs

Material	S_{BET} (m²/g)	P (bar)	T (K)	CO_2 uptake (mmol/g)	References
COFs					
COF-1	750	55	298	5.2	[11]
COF-5	1,670	55	298	19.8	[11]
COF-6	750	55	298	7.0	[11]
COF-8	1,350	55	298	14.3	[11]
COF-10	1,760	55	298	23.0	[11]
COF-102	3,620	55	298	27.3	[11]
COF-103	3,530	55	298	27.0	[11]
		30		10.6	
CMPs					
CMP-1	837	1	298	1.18	[55]
			273	2.05	
CMP-1-(CH₃)₂	899	1	298	0.94	[55]
			273	1.64	
CMP-1-(OH)₂	1,043	1	298	1.07	[55]
			273	1.80	
CMP-1-NH₂	710	1	298	0.95	[55]
			273	1.64	
CMP-1-COOH	522	1	298	0.95	[55]
			273	1.60	
PAF-1	5,600	40	298	29.5	[30]
		1	298	2.05	[29]
			273	1.09	
PAF-3	2,932	1	273	3.48	[29]
			298	1.81	
PAF-4	2,246	1	273	2.41	[29]
			298	1.16	
PPF-1	1,740	1	273	6.07	[44]
PPF-2	1,470	1	273	5.55	
PPN-1	1,249	60	295	11	[48]
PPN-2	1,764	60	295	19	[48]
PPN-3	2,840	60	295	25.3	[48]
PPN-4	6,461	50	295	48.2	[32]
PPN-6-SO₃H	1,254	1	295	3.6	[56]
PPN-6-SO₃-Li	1,186	1	295	3.7	[56]
PPN-6-SO₃NH₄	593	1	295	3.7	[57]
PPN-6-CH₂DETA	555	1	295	4.3	[58]

incorporated into CMP materials (Fig. 4.8a). N_2 sorption results indicate their surface areas range from 522 to 1,043 m² g⁻¹ (Fig. 4.8b). Non-functionalized CMP-1 network showed the highest CO_2 uptake with a value of 1.18 mmol g⁻¹, at 298 K and 1 bar. CMP-1-(OH)₂ adsorbed less CO_2 (1.07 mmol g⁻¹, at 298 K

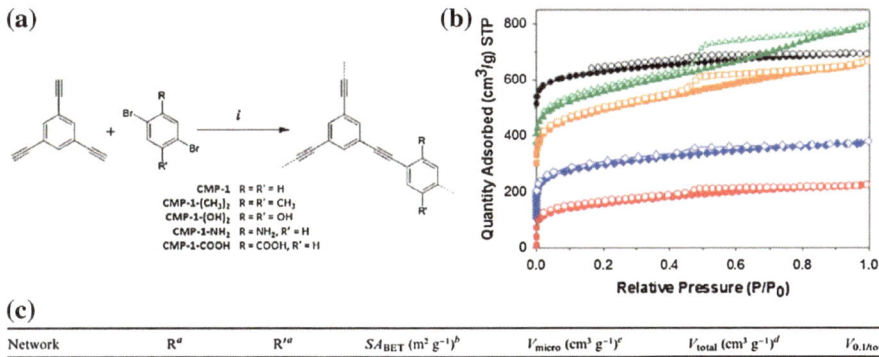

Fig. 4.8 **a** Synthesis of functionalised CMPs using *i* DMF, NEt₃, Pd(PPh₃)₄, CuI, 100 °C, 72 h; **b** nitrogen adsorption (closed)/desorption (open) isotherms (77 K) for CMP-1 (*black*), CMP-1-(CH₃)₂ (*green*), CMP-1-(OH)₂ (*orange*), CMP-1-NH₂ (*blue*) and CMP-1-COOH (*red*) each off-set by 100 cm³ g⁻¹ for clarity; **c** surface areas and pore volumes for CMP networks. Reprinted with permission from Ref. [55]. Copyright 2011, Royal Society of Chemistry

and 1 bar), despite exhibiting a higher surface area and pore volume. The dimethyl network, CMP-1-(CH₃)₂, showed the lowest uptake of CO_2 (0.94 mmol g⁻¹), despite having higher surface area than CMP-1 (Fig. 4.8c). In addition, the isosteric heat of adsorption (Qst) of gas molecules for porous materials is an important indicator to assess the affinity of guest molecules to host materials. The carboxylic acid functionalized network indicates the highest isosteric heat of sorption for CO_2, supporting recent computational predictions for metal–organic frameworks that carboxylic acid group is favorable to enhance the interaction between porous materials and CO_2 molecules. The result suggests that acid functionalized frameworks could be widely studied in CO_2 capture and separation application.

4.4.2 Tuning the Inner Surface of POFs Using Post-Modification Strategy

PAF-1 displays high BET surface area with a value of 5,600 m² g⁻¹, narrow pore size distribution centered at 1.41 nm, high-thermal stability (decomposing temperature up to 520 °C indicated by TGA, and high water stability. Therefore, PAF-1 is an excellent candidate as the starting materials for post-modification. Zhou et al. reported two methods to tune the surface features of the PAF-1's wall. One is the synthesis of sulfonate-grafted network (Fig. 4.9a) and the other is the synthesis of amine-grafted wall (Fig. 4.9b) [56–58]. When the reaction occurs, the

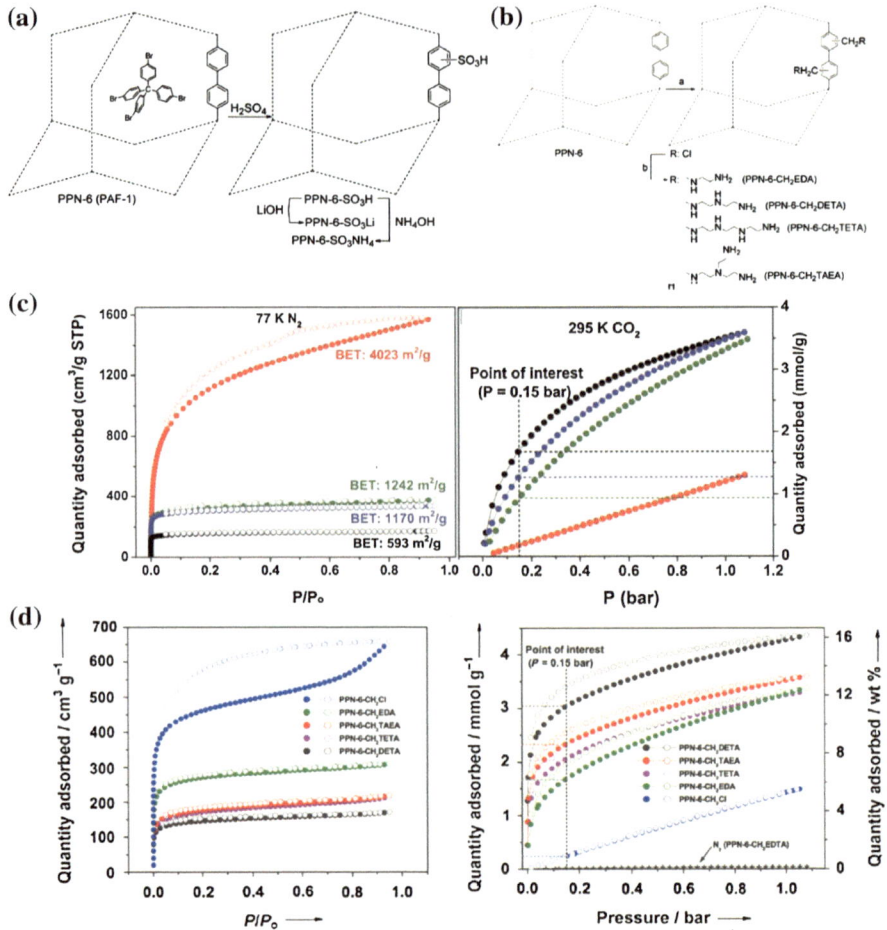

Fig. 4.9 **a** Synthetic route for sulfonate functionalized PPNs; **b** synthetic route for amine function-alized PPNs; **c** 77 K N_2 sorption isotherms and 295 K CO_2 adsorption isotherms. PPN-6 (*red*), PPN-6-SO$_3$H (*green*), PPN-6-SO$_3$Li (*blue*), and PPN-6-SO$_3$NH$_4$ (*black*); **d** N_2 adsorption and desorption isotherms at 77 K and CO_2 adsorption and desorption isotherms, as well as PPN-6-CH$_2$DETA N_2 adsorption, at 295 K. Reprinted with permission from Ref. [57]. Copyright 2013, Royal Society of Chemistry. Reprinted with permission from Ref. [58]. Copyright 2012, Wiley-VCH

surface area of resulting product sharply decreases compared with the starting materials (Fig. 4.9c, d). Excitingly, both the sulfonate-grafted and amine-grafted materials show higher CO_2 uptake than that of PAF-1. At 295 K and 1 bar, PPN-6-CH$_2$DETA exhibits exceptionally high CO_2 uptake with a value of 4.3 mmol g^{-1} (15.8 wt%), which is the highest among the microporous organic polymers reported so far.

Our group also reported a post-modified method to change the properties of PAF materials. PAF-18 [33] and PAF-26 [51] were synthesized using Sonogashira–Hagihara cross-coupling reaction. Through introduction of functional groups into

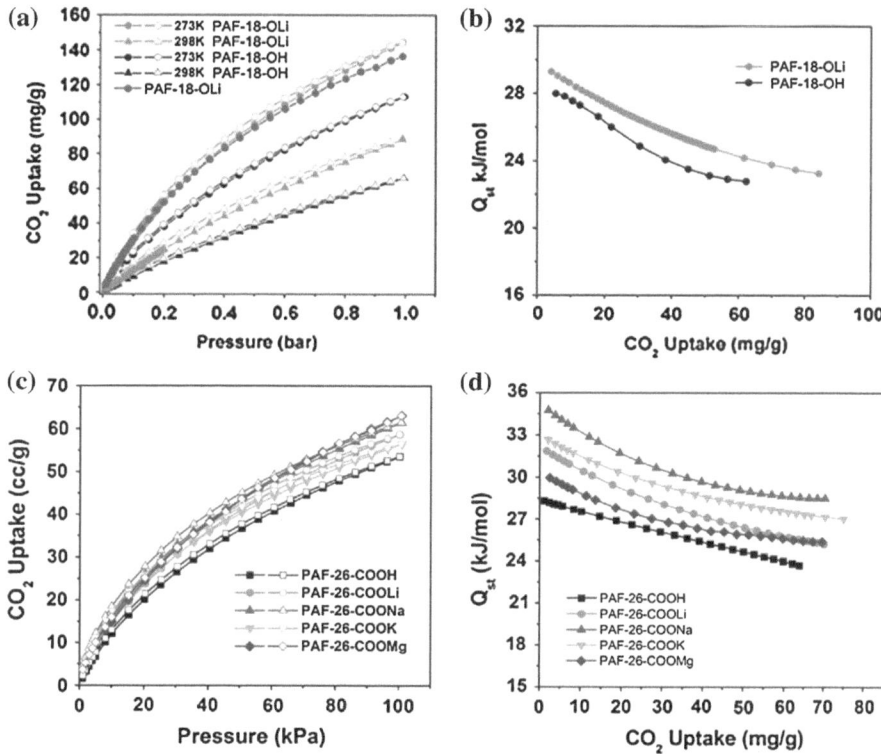

Fig. 4.10 **a** CO_2 adsorption and desorption isotherms of PAF-18-OH and PAF-18-OLi at 273 and 298 K. CO_2 adsorption of PAF-18-OLi obtained after 10 days exposure in humid air (*blue circles*) at 273 K; **b** plots of the isosteric heat of adsorption (Qst) for CO_2 of PAF-18-OH and PAF-18-OLi; **c** gas sorption isotherms of CO_2 for PAF-26-COOH, PAF-26-COOLi, PAF-26-COONa, PAF-26-COOK and PAF-26-COOMg at 273 K and 101 kPa; **d** the isosteric heats of adsorption for CO_2 of PAF-26-COOH, PAF-26-COOLi, PAF-26-COONa, PAF-26-COOK and PAF-26-COOMg as a function of gas uptake

PAF skeletons, their corresponding post-modified products are obtained. Although the surface area of resulted product decreased, the CO_2 uptakes enhanced (Fig. 4.10a, c). Compared with the starting materials, the stronger interaction between CO_2 molecules and the host materials (PAF-18-Li and PAF-26-COOM) is verified by their higher isosteric heats of adsorption (Fig. 4.10b, d), which is due to the higher polarity of metal ions than that of protons and the small-pore effect.

4.4.3 Carbonization of POFs

POFs are composed of light elements such as C, H, O, N, B, etc. Therefore, they are excellent precursors for preparing nanoporous carbon materials via high temperature decomposition. Qiu et al. reported a series of carbonized PAF-1 s under

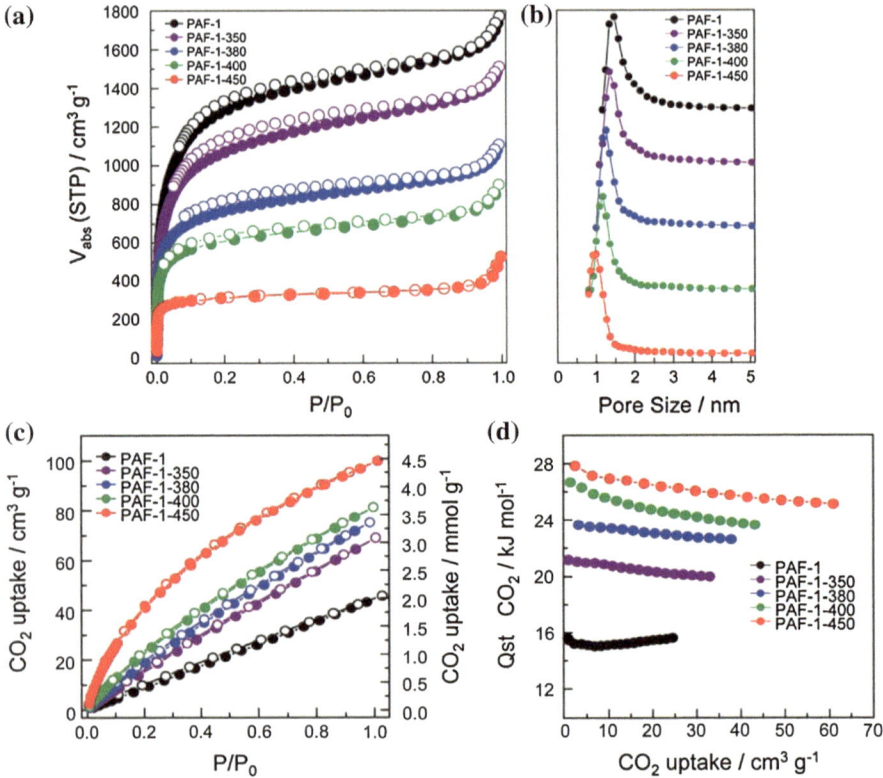

Fig. 4.11 a N_2 sorption isotherms of PAF-1 and PAF-1-350, PAF-1-380, PAF-1-400, PAF-1-450; **b** pore size distributions of PAF-1 and PAF-1-350, PAF-1-380, PAF-1-400, PAF-1-450 derived from N_2 adsorption calculated by Density Functional Theory (DFT); **c** CO_2 adsorption and desorption isotherms of PAF-1 and carbonized samples at 273 K; **d** Q_{stCO_2} of PAF-1 and carbonized samples as a function of the amount of CO_2 adsorbed. Reprinted with permission from Ref. [59]. Copyright 2013, Royal Society of Chemistry

different temperature from 350 to 450 °C [59]. Calculated from the N_2 sorption results using BET model, the surface areas of PAF-1-350, PAF-1-380, PAF-1-400, and PAF-1-450 were 4,033, 2,881, 2,292, and 1,191 m^2 g^{-1}, respectively (Fig. 4.11a). Calculated by the density functional theory (DFT) method, the total pore volumes of these samples dropped from 2.43 cm^3 g^{-1} of PAF-1 to 0.53 cm^3 g^{-1} of PAF-1-450. Meanwhile, the pore size distribution shrunk from 1.44 nm of PAF-1 to 1.00 nm of PAF-1-450 (Fig. 4.11b). Especially, the CO_2 uptake of PAF-1-450 reached 4.5 mmol g^{-1}, at 273 K and 1 bar (Fig. 4.11c). As expected, the Qst CO_2 of carbonized PAF-1 increased evidently by comparison with the original PAF-1 (Fig. 4.11d).

CCS is a hot topic due to its important environmental sustainability which concerns our daily life. POFs display some advantages in CO_2 capture thanks to their high surface area, high stability, and adjustable structure. The pioneering studies provide guidelines for the construction of promising POFs for CCS.

4.5 Small Hydrocarbon Storage

Small hydrocarbons (C_1, C_2, and C_3) are important energy sources and chemical feedstock in the petrochemical industry. In order to fully utilize these light hydrocarbons, it is essential to have high quality and purity of such basic chemicals; thus, separations of these light hydrocarbons are important industrial processes. In general, to obtain high-purity single-component gas product, purification of the small hydrocarbons is achieved using cryogenic distillation, which is a high energy consumption process.

The pressure swing adsorption (PSA) process is a cost-effective and highly efficient technique for separating small hydrocarbons [60–62]. In the PSA process, the separation efficiency basically depends on the adsorption properties of solid adsorbents. Thus, extensive efforts have been made to expand efficient solid adsorbents for PSA technology. Typically, PSA is carried out under kinetic dynamic conditions and run a cycle in several minutes. To solve this challenge, the adsorbent must possess fast separation ability. Porous solid materials, such as zeolites, and activated carbons have been examined for storage and separations of these light hydrocarbons. However, these traditional adsorbents exhibit a low separation capacity and selectivity in the process of separating small hydrocarbons mixtures. Therefore, development of new adsorbents is required. MOFs have been exploited as a new type of sorbents for hydrocarbon separation [63–66]. Compared with MOFs, the new family of porous material, POFs have been documented as new carriers for gas capture and storage [1–3]. Because of their controllable pore connectivity, chemical tailorability, high stability, and large adsorption capacity, POFs have emerged as appropriate candidates for small hydrocarbons storage (Table 4.5). Compared with utilization of POFs in the field of H_2, CO_2, and CH_4 storages, studies on small hydrocarbons are relatively rare.

Mercouri et al. reported the synthesis of mesoporous polymeric organic frameworks (mesoPOF) using surfactant mediated polymerization of phlorglucinol (1,3,5-trihydrox-ybenzene) and terephthalaldehyde under solvothermal conditions (Fig. 4.12a) [67]. The mesoPOFs present high surface areas up to 1,000 $m^2\ g^{-1}$, and their pore sizes range from micropores to large mesopores depending on the amount

Table 4.5 Small hydrocarbons storage in POFs

Materials	S_{BET} (m^2/g)	T (K)	P (bar)	CH_4 uptake (cm^3/g)	C_2H_4 uptake (cm^3/g)	C_2H_6 uptake (cm^3/g)	C_3H_8 uptake (cm^3/g)	References
MesPOF-1	1,027	273	1	~18		57		[67]
MesPOF-1	902			~9		~33		[67]
MesPOF-1	548			~7		~24		[67]
Zn-CTF-400	1,411	298	1	11		70	112	[68]
Zn-CTF-500	1,848	298	1	15		90	161	[68]
Zn-CTF-600	1,331	298	1	9		54	103	[68]
MCOF-1	874	298	1	9	36	44	55	[69]

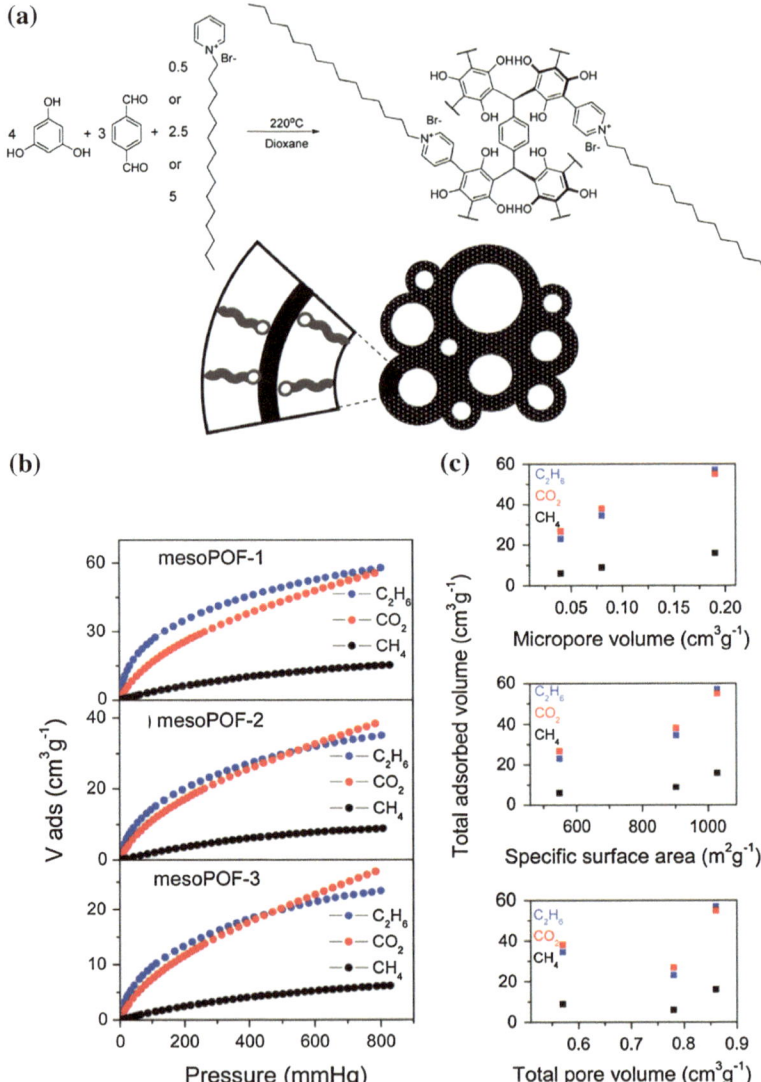

Fig. 4.12 **a** Polymerization reaction between phloroglucinol and terephthaladehyde, the covalent linking of CPyBr, and the subsequent formation of the framework; **b** adsorption isotherms for C_2H_6, CO_2, and CH_4 for mesoPOF-1, mesoPOF-2, and mesoPOF-3 at 273 K; **c** the total adsorbed volume of C_2H_6, CO_2, and CH_4 of each mesoPOF versus micropore volume, specific surface area, and total pore volume. Reprinted with permission from Ref. [67]. Copyright 2012, American Chemical Society

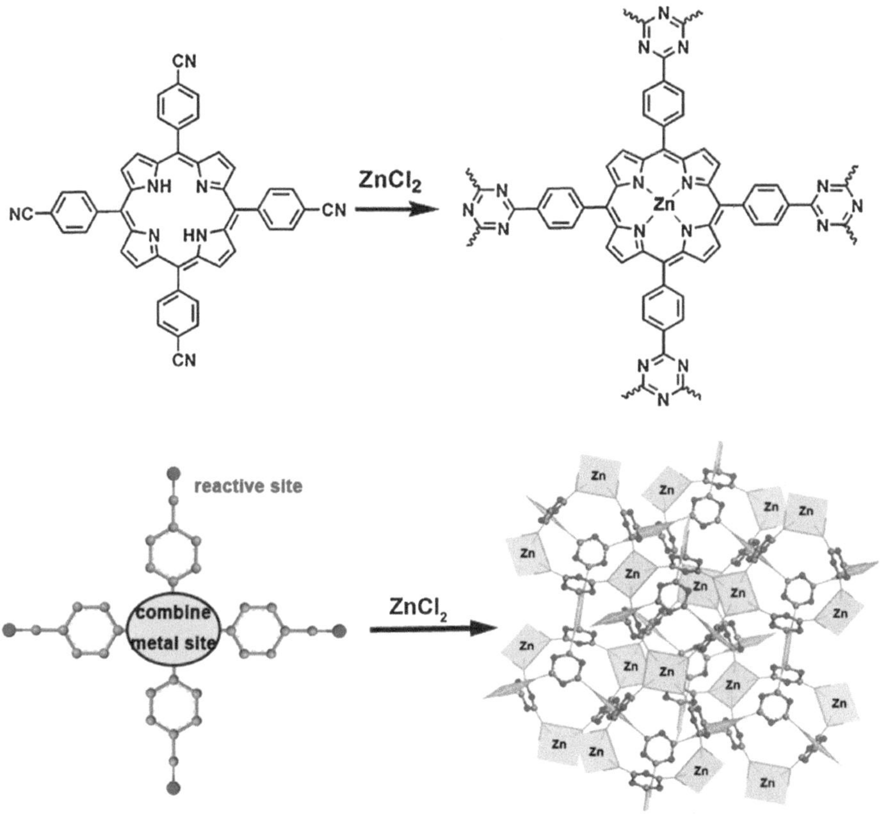

Fig. 4.13 Schematic representation of the synthesis of ZnP-CTFs

of surfactant used. MesoPOF-1 exhibits high adsorption capacity of C_2H_6 with value of 57 cm^3 g^{-1}, at 1 bar and 273 K (Fig. 4.12b). MesoPOF-2 and mesoPOF-3 show lower uptakes for all gases than that of mesoPOF-1.

Recently, we successfully synthesized novel ZnP-CTF materials through ionothermal reaction (Fig. 4.13) [68]. By controlling synthetic temperature, tunable surface areas, and pore sizes of ZnP-CTF materials are achieved. More valuably, ZnP-CTF materials show different adsorption capacities for small hydrocarbons according to the number of carbon atoms (Fig. 4.14). At 298 K and 101 kPa, the C_3H_8 uptake for ZnP-CTF-400, ZnP-CTF-500, and ZnP-CTF-600 is 112, 161, and 103 cm^3 g^{-1}, respectively. The C_2H_6 uptake is 70 cm^3 g^{-1} for ZnP-CTF-400, 90 cm^3 g^{-1} for ZnP-CTF-500, and 54 cm^3 g^{-1} for ZnP-CTF-600 under the same condition. Furthermore, we explored the utility of ZnP-CTF-400 for high performance GC separation of small hydrocarbons mixture owing to their

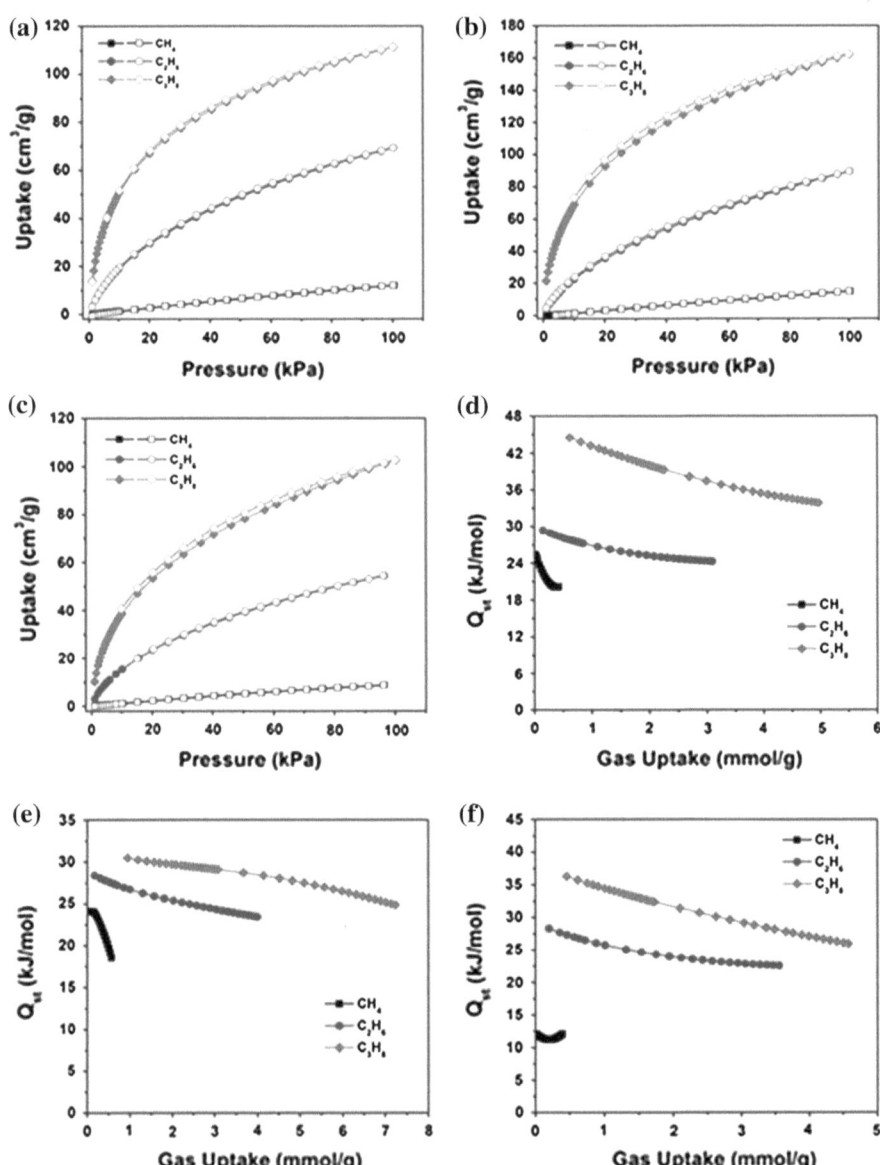

Fig. 4.14 Gas uptake isotherms for **a** ZnP-CTF-400, **b** ZnP-CTF-500 and **c** ZnP-CTF-600 at 298 K, 101 kPa (*black* CH_4, *red* C_2H_6, and *olive* C_3H_8); The isosteric heats of adsorption for ZnP-CTF-400 (**d**) ZnP-CTF-500 (**e**) and ZnP-CTF-600 (**f**) in function of the gas uptake (*black* for CH_4, *red* for C_2H_6 and *olive* C_3H_8)

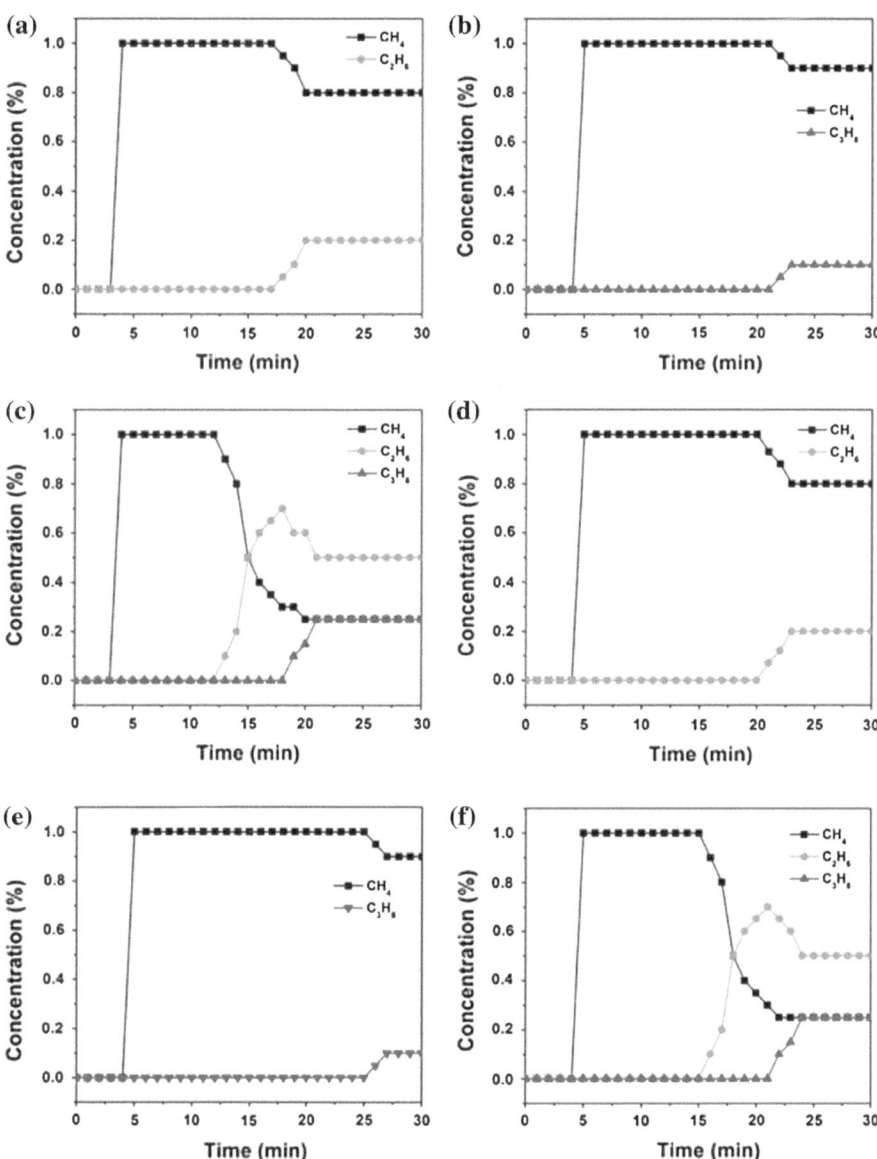

Fig. 4.15 Breakthrough curves of **a** C_2H_6/CH_4 mixture [20:80 (vol)]; **b** C_3H_8/CH_4 mixture [10:90 (vol)]; **c** $C_3H_8/C_2H_6/CH_4$ mixture [25:50:25 (vol)] for ZnP-CTF-400 and **d** C_2H_6/CH_4 mixture [20:80 (vol)]; **e** C_3H_8/CH_4 mixture [10:90 (vol)]; **f** $C_3H_8/C_2H_6/CH_4$ mixture [25:50:25 (vol)] for ZnP-CTF-500. These were measured at 298 K, 3 Mpa

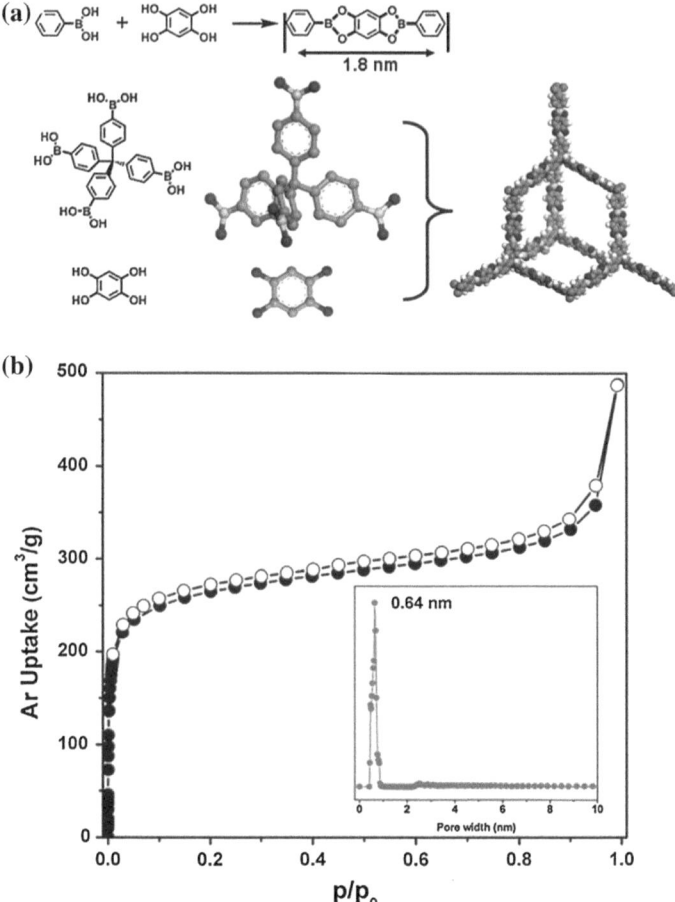

Fig. 4.16 **a** Schematic representation of the synthesis of MCOF-1; **b** Argon adsorption and desorption (*open symbols*) isotherm of MCOF-1 measured at 87 K. *Inset* Pore size distribution calculated by the NLDFT model

different polarizability. The fast dynamic breakthrough tests further proved that the ZnP-CTF-400 and ZnP-CTF-500 could separate small hydrocarbons based on their hierarchical isosteric heats in an actual adsorption-based separation process (Fig. 4.15).

We have also successfully synthesized a new microporous crystalline material (MCOF-1) with narrow pore size distribution of 0.64 nm (Fig. 4.16) [69]. MCOF-1 possessed architectural stability and porosity after evacuating solvent molecules. Additionally, MCOF-1 showed high adsorption capacity for C_2H_4, C_2H_6, and C_3H_8. At 298 K and 1 bar, the uptake of MCOF-1 is 9, 36, 44 and

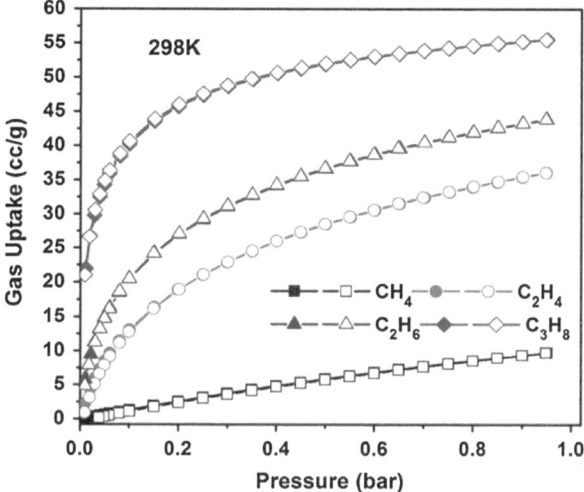

Fig. 4.17 Gas (*black* CH$_4$, *red* C$_2$H$_4$, *blue* C$_2$H$_6$ and *olive* C$_3$H$_8$) uptake isotherms of MCOF-1 at 298 K

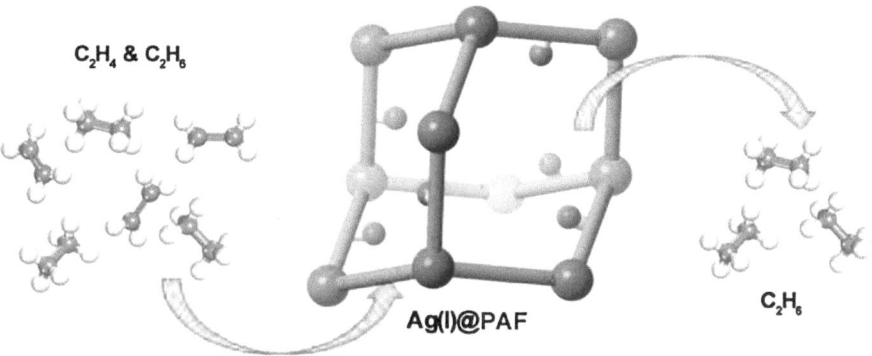

Fig. 4.18 Ag(I) incorporated PAF-1 for highly selective adsorption of ethylene over ethane. Reprinted with permission from Ref. [70]. Copyright 2014, American Chemical Society

55 cm^3 g^{-1} for CH$_4$, C$_2$H$_4$, C$_2$H$_6$, and C$_3$H$_8$, respectively (Fig. 4.17). More valuably, MCOF-1 exhibited significantly high selectivity for C$_3$H$_8$ over CH$_4$ and high selectivities for C$_2$H$_6$ and C$_2$H$_4$ over CH$_4$.

As reported, Cu(I) and Ag(I) ions can form π-complexation with the carbon–carbon double bonds of olefin molecules in solutions. If such kinds of π-complexation could be introduced to POP skeletons, the π-complexation will afford strong interactions between the ethylene molecules and the framework.

Fig. 4.19 a N$_2$ sorption
isotherms at 77 K for PAF-1
(*black*), PAF-1-SO$_3$H (*red*),
and PAF-1-SO$_3$Ag (*blue*);
b C$_2$H$_4$ (*black*) and C$_2$H$_6$
(*red*) sorption isotherms of
PAF-1-SO$_3$Ag at 296 K.
Reprinted with permission
from Ref. [70]. Copyright
2014, American Chemical
Society

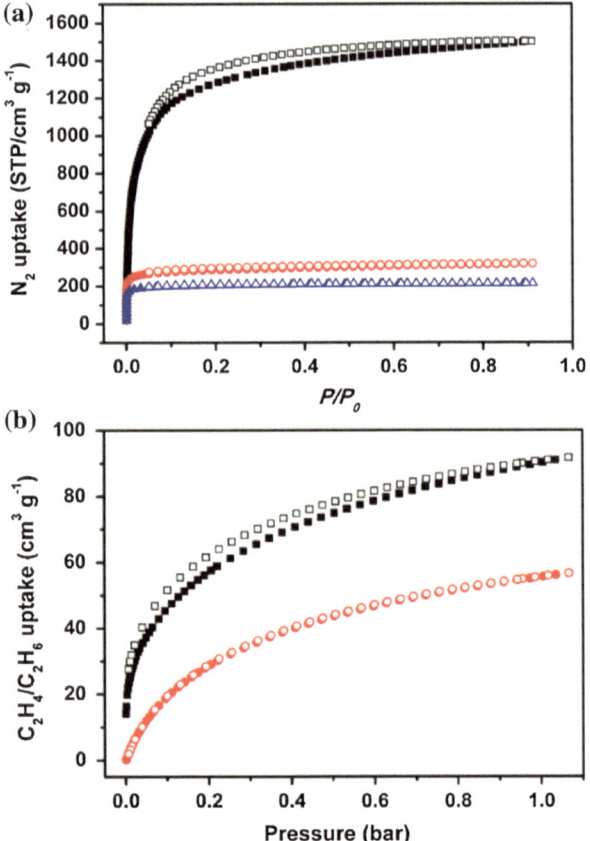

Recently, Ma's group first introduced π-complexation to PAF-1 [70]. By Ag(I)
ion exchange of sulfonate-grafted PAF-1 (denoted PAF-1-SO$_3$H), Ag(I) ions were
decorated into PAF-1 networks to form PAF-1-SO$_3$Ag (Fig. 4.18). N$_2$ gas sorption isotherms at 77 K reveal that BET surface areas of PAF-1 and PAF-1-SO$_3$Ag
are 4,714 and 783 m^2 g^{-1}, respectively (Fig. 4.19a). Pore size distribution analysis
indicates that its pore size is reduced from ~15 Å of PAF-1 to ~8 Å for PAF-
1-SO$_3$Ag. At 296 K and 1 atm, the resulting product affords the unexceptional
increase in ethylene uptake, from 57 cm^3 g^{-1} of PAF-1 to 91 cm^3 g^{-1} of PAF-1-
SO$_3$Ag (Fig. 4.19b). They also adopt transient breakthrough experiments to separate the mixture. As a result, PAF-1-SO$_3$Ag could produce 99.95 % + pure C$_2$H$_4$
in a PSA operation.

Small hydrocarbons storage in POFs is an emerging research area. However,
current studies show that POFs are promising candidates to stratify this requirement.

References

1. Dawson R, Cooper A, Adams D (2012) Nanoporous organic polymer networks. Prog Polym Sci 37:530–563
2. Xiang Z, Cao D (2013) Porous covalent-organic materials: synthesis, clean energy application and design. J Mater Chem A 1:2691–2718
3. Kalidindi S, Fischer R (2013) Covalent organic frameworks and their metal nanoparticle composites: prospects for hydrogen storage. Phys Status Solidi B 250:1119–1127
4. IPCC (2007) Summary for policymakers. In: Climate change 2007: the physical science basis. Cambridge University Press, Cambridge
5. Kim M, Bae Y, Choi D et al (2006) Kinetic separation of landfill gas by a two-bed pressure swing adsorption process packed with carbon molecular sieve: nonisothermal operation. Ind Eng Chem Res 45:5050–5058
6. Babarao R, Hu Z, Jiang J et al (2007) Storage and separation of CO_2 and CH_4 in silicalite, C168 schwarzite, and IRMOF-1: a comparative study from Monte Carlo simulation. Langmuir 23:659–666
7. Li J, Sculley J, Zhou H (2012) Metal–organic frameworks for separations. Chem Rev 112:869–932
8. McKeown N, Budd P (2010) Exploitation of intrinsic microporosity in polymer-based materials. Macromolecules 43:5163–5176
9. Zou X, Ren H, Zhu G (2013) Topology-directed design of porous organic frameworks and their advanced applications. Chem Commun 49:3925–3936
10. http://www1.eere.energy.gov/hydrogenandfuelcells/storage/currenttechnology.html
11. Furukawa H, Yaghi O (2009) Storage of hydrogen, methane, and carbon dioxide in highly porous covalent organic frameworks for clean energy applications. J Am Chem Soc 131:8875–8883
12. Li Y, Yang R (2008) Hydrogen storage in metal-organic and covalent-organic frameworks by spillover. AlChE J 54:269–279
13. Tilford R, Mugavero S, Pellechia P et al (2008) Tailoring microporosity in covalent organic frameworks. Adv Mater 20:2741–2746
14. Rabbani M, Sekizkardes A, Kahveci Z et al (2013) A 2D mesoporous imine-linked covalent organic framework for high pressure gas storage applications. Chem Eur J 19:3324–3328
15. Kahveci Z, Islamoglu T, Shar G et al (2013) Targeted synthesis of a mesoporous triptycene-derived covalent organic framework. CrystEngComm 15:1524–1527
16. Song J, Sun J, Liu J et al (2014) Thermally/hydrolytically stable covalent organic frameworks from a rigid macrocyclic host. Chem Commun 50:788–791
17. McKeown N, Budd P, Book D (2007) Microporous polymers as potential hydrogen storage materials. Macromol Rapid Commun 28:995–1002
18. Ghanem B, Msayib K, McKeown N et al (2007) A triptycene-based polymer of intrinsic microposity that displays enhanced surface area and hydrogen adsorption. Chem Commun 67–69
19. Makhseed S, Samuel J (2008) Hydrogen adsorption in microporous organic framework polymer. Chem Commun 4342–4344
20. Chen Q, Luo M, Hammershøj et al (2012) Microporous polycarbazole with high specific surface area for gas storage and separation. J Am Chem Soc 134:6084–6087
21. Chen Q, Liu D, Luo M et al (2014) Nitrogen-containing microporous conjugated polymers via carbazole-based oxidative coupling polymerization: preparation, porosity, and gas uptake. Small 10:308–315
22. Rose M, Böhlmann W, Sabo M et al (2008) Element–organic frameworks with high permanent porosity, Chem Commun 2462–2464
23. Fritsch J, Rose M, Wollmann P et al (2010) New element organic frameworks based on Sn, Sb, and Bi, with permanent porosity and high catalytic activity. Materials 3:2447–2462

24. Jiang J, Su F, Trewin A et al (2008) Synthetic control of the pore dimension and surface area in conjugated microporous polymer and copolymer networks. J Am Chem Soc 130:7710–7720
25. Hasell T, Wood C, Clowes R et al (2009) Palladium nanoparticle incorporation in conjugated microporous polymers by supercritical fluid processing. Chem Mater 22:557–564
26. Li A, Lu R, Wang Y et al (2010) Lithium-doped conjugated microporous polymers for reversible hydrogen storage. Angew Chem Int Ed 49:3330–3333
27. Reich T, Jackson K, Li S et al (2011) Synthesis and characterization of highly porous borazine-linked polymers and their performance in hydrogen storage application. J Mater Chem 21:10629–10632
28. Jackson K, Reich T, El-Kaderi H (2012) Targeted synthesis of a porous borazine-linked covalent organic framework. Chem Commun 48:8823–8825
29. Ben T, Pei C, Zhang D et al (2011) Gas storage in porous aromatic frameworks (PAFs). Energy Environ Sci 4:3991–3999
30. Ben T, Ren H, Ma S et al (2009) Targeted synthesis of a porous aromatic framework with high stability and exceptionally high surface area. Angew Chem Int Ed 121:9621–9624
31. Konstas K, Taylor J, Thornton A et al (2012) Lithiated porous aromatic frameworks with exceptional gas storage capacity. Angew Chem Int Ed 124:6743–6746
32. Yuan D, Lu W, Zhao D et al (2011) Highly stable porous polymer networks with exceptionally high gas-uptake capacities. Adv Mater 23:3723–3725
33. Ma H, Ren H, Zou X et al (2013) Novel lithium-loaded porous aromatic framework for efficient CO_2 and H_2 uptake. J Mater Chem A 1:752–758
34. Yuan S, Dorney B, White D et al (2010) Microporous polyphenylenes with tunable pore size for hydrogen storage. Chem Commun 46:4547–4549
35. Rabbani M, El-Kaderi H (2012) Synthesis and characterization of porous benzimidazole-linked polymers and their performance in small gas storage and selective uptake. Chem Mater 24:1511–1517
36. Lee J, Wood C, Bradshaw D et al (2006) Hydrogen adsorption in microporous hypercrosslinked polymers. Chem Commun 2670–2672
37. Wood C, Tan B, Trewin A et al (2007) Hydrogen storage in microporous hypercrosslinked organic polymer networks. Chem Mater 19:2034–2048
38. Germain J, Fréchet J, Svec F et al (2007) Hypercrosslinked polyanilines with nanoporous structure and high surface area: potential adsorbents for hydrogen storage. J Mater Chem 17:4989–4997
39. Germain J, Svec F, Fréchet J (2008) Preparation of size-selective nanoporous polymer networks of aromatic rings: potential adsorbents for hydrogen storage. Chem Mater 20:7069–7076
40. Mendoza-Cortés J, Han S, Goddard W III (2012) High H_2 uptake in Li-, Na-, K-metalated covalent organic frameworks and metal organic frameworks at 298 K. J Phys Chem A 116:1621–1631
41. Han S, Mendoza-Cortés J, Goddard Iii W et al (2009) Recent advances on simulation and theory of hydrogen storage in metal–organic frameworks and covalent organic frameworks. Chem Rev 38:1460–1476
42. Mendoza-Cortes J, Goddard W III, Furukawa H et al (2012) A covalent organic framework that exceeds the DOE 2015 volumetric target for H_2 uptake at 298 K. J Phys Chem Lett 3:2671–2675
43. Huang L, Zeng X, Cao D (2014) Tetrahedral node diamondyne frameworks for CO_2 adsorption and separation. J Mater Chem A 2:4899–4902
44. Zhu Y, Long H, Zhang W (2013) Imine-linked porous polymer frameworks with high small gas (H_2, CO_2, CH_4, C_2H_2) uptake and CO_2/N_2 selectivity. Chem Mater 25:1630–1635
45. Li L, Ren H, Yuan Y et al (2014) Construction and adsorption properties of porous aromatic frameworks via AlCl3-triggered coupling polymerization. J Mater Chem A 2:11091–11098
46. Rabbani M, Reich T, Kassab R et al (2012) High CO_2 uptake and selectivity by triptycene-derived benzimidazole-linked polymers. Chem Commun 48:1141–1143

47. Wood C, Tan B, Trewin A et al (2008) Microporous organic polymers for methane storage. Adv Mater 20:1916–1921
48. Lu W, Yuan D, Zhao D et al (2010) Porous polymer networks: synthesis, porosity, and applications in gas storage/separation. Chem Mater 22:5964–5972
49. Arab P, Rabbani M, Sekizkardes A et al (2014) Copper (I)-catalyzed synthesis of nanoporous azo-linked polymers: impact of textural properties on gas storage and selective carbon dioxide capture. Chem Mater 6:1385–1392
50. Mendoza-Cortés J, Han S, Furukawa H et al (2010) Adsorption mechanism and uptake of methane in covalent organic frameworks: theory and experiment. J Phys Chem A 114:10824–10833
51. Ma H, Ren H, Zou X et al (2014) Post-metalation of porous aromatic frameworks for highly efficient carbon capture from $CO_2 + N_2$ and $CH_4 + N_2$ mixtures. Polym Chem 5:144–152
52. Sumida K, Rogow D, Mason J et al (2012) Carbon dioxide capture in metal–organic frameworks. Chem Rev 112:724–781
53. Liu J, Thallapally P, McGrail B et al (2012) Progress in adsorption-based CO_2 capture by metal–organic frameworks. Chem Soc Rev 41:2308–2322
54. Rochelle G (2009) Amine Scrubbing for CO_2 Capture. Science 325:1652–1654
55. Dawson R, Adams D, Cooper A (2011) Chemical tuning of CO_2 sorption in robust nanoporous organic polymers. Chem Sci 2:1173–1177
56. Lu W, Yuan D, Sculley J et al (2011) Sulfonate-grafted porous polymer networks for preferential CO_2 adsorption at low pressure. J Am Chem Soc 133:18126–18129
57. Lu W, Verdegaal W, Yu J et al (2013) Building multiple adsorption sites in porous polymer networks for carbon capture applications. Energy Environ Sci 6:3559–3564
58. Lu W, Sculley J, Yuan D et al (2012) Polyamine-tethered porous polymer networks for carbon dioxide capture from flue gas. Angew Chem Int Ed 51:7480–7484
59. Ben T, Li Y, Zhu L et al (2012) Selective adsorption of carbon dioxide by carbonized porous aromatic framework (PAF). Energy Environ Sci 5:8370–8376
60. Magnowski N, Avila A, Lin C et al (2011) Extraction of ethane from natural gas by adsorption on modified ETS-10. Chem Eng Sci 66:1697–1701
61. Arruebo M, Coronas J, Menendez M et al (2001) Separation of hydrocarbons from natural gas using silicalite membranes. Sep Purif Tech 25:275–286
62. Triebe R, Tezel W, Khulbe K et al (1996) Adsorption of methane, ethane and ethylene on molecular sieve zeolites. Gas Sep Purif 10:81–84
63. He Y et al (2012) A robust doubly interpenetrated metal–organic framework constructed from a novel aromatic tricarboxylate for highly selective separation of small hydrocarbons. Chem Commun 48:6493–6495
64. He Y, Krishna R, Chen B et al (2012) Metal–organic frameworks with potential for energy-efficient adsorptive separation of light hydrocarbons. Energy Environ Sci 5:9107–9120
65. Horike S, Inubushi Y, Hori T et al (2012) A solid solution approach to 2D coordination polymers for CH_4/CO_2 and CH_4/C_2H_6 gas separation: equilibrium and kinetic studies. Chem Sci 3:116–120
66. Duan J et al (2013) High CO_2/CH_4 and C_2 hydrocarbons/CH_4 selectivity in a chemically robust porous coordination polymer. Adv Funct Mater 23:3525–3530
67. Katsoulidis A, Kanatzidis M et al (2012) Mesoporous hydrophobic polymeric organic frameworks with bound surfactants, selective adsorption of C_2H_6 versus CH_4. Chem Mater 24:471–479
68. Ma H, Ren H, Meng S et al (2013) Novel porphyrinic porous organic frameworks for high performance separation of small hydrocarbons. Sci Rep 3:2611–2617
69. Ma H, Ren H, Meng S et al (2013) A 3D microporous covalent organic framework with exceedingly high C_3H_8/CH_4 and C_2 hydrocarbon/CH_4 selectivity. Chem Commun 49:9773–9775
70. Li B, Zhang Y, Krishna R et al (2014) Introduction of π-complexation into porous aromatic framework for highly selective adsorption of ethylene over ethane. J Am Chem Soc 36:8654–8660

Chapter 5
Porous Organic Frameworks for Catalysis

Abstract POFs possess high stability, high surface area, tuned pore size, adjustable skeletons, and etc. POFs provide the versatile platforms for catalytic applications. Through the pre-synthetic and post-synthetic strategies, POFs with catalytic performance could be achieved. In this chapter, we introduce two types of catalysts. One is POFs with metal-containing site as catalysts and the other is special organic groups as heterogeneous catalyst. Furthermore, POFs could be served as catalyst support owing to their high surface area and various coordinating sites. The typical strategies for the synthesis of catalytic POFs are also presented.

Keywords Pre-synthetic and post-synthetic strategies · Metal-containing sites · Catalyst support · Metal-free catalyst · Heterogeneous catalyst

5.1 Introduction

It has been documented that POFs possess various porous structures, high stability, and adjustable skeletons. Thus a multipurpose platform has been established of which POFs have been employed for catalytic applications [1, 2]. Commonly, there are two promising strategies to achieve this requirement (Fig. 5.1). One is directly synthesis of catalytic POFs by introducing building blocks containing active catalytic groups; the other is via post-synthetic method, of which the active catalyst is combined with the original POF materials by post-synthetic treatment. Based on the types of active catalyst, two major classes will be introduced: (1) POFs t with metal-containing site as catalysts; (2) POFs with incorporate rigid organic building blocks as heterogeneous catalysts. In this chapter, we will present the strategy for constructing catalytic POFs and their applications.

© The Author(s) 2015   87
G. Zhu and H. Ren, *Porous Organic Frameworks*, SpringerBriefs in Green Chemistry for Sustainability, DOI 10.1007/978-3-662-45456-5_5

Fig. 5.1 The promising methods for constructing POFs with catalytic property. Reprinted with permission from Ref. [1]. Copyright 2011, American Chemical Society

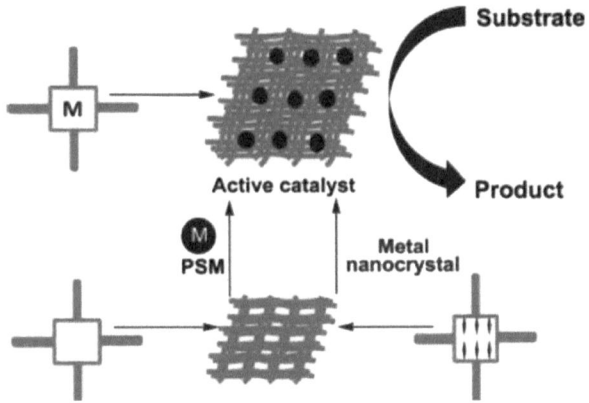

5.2 POFs with Metal-Containing Sites for Catalysis

The catalytic properties of catalysts are determined by the types of the active sites. Many studies have illustrated the strategies to construct porous materials with catalytic properties. Thus the design and synthesis of catalytic POFs could be directed according to the previous investigations. When we prepare catalytic POFs, first, the general catalytic reaction is confirmed, and then we need to establish reasonable procedure to fulfill the targeted aims. It is notable that the active site of POFs could be achieved by de novo or post-synthetic means.

5.2.1 POFs with Metal-Containing Sites for Catalysis

Incorporation of the required metal in the original building units is an easy and direct method. The 2,2-Bipyridine, porphyrin ring, and phthalocyanines ring are interesting powerful ligand with rich coordination chemistry in the square-planar coordination site. Thus they provide potential for POFs to fulfill the demands for the incorporation of the metals (Fig. 5.2).

Jiang et al. described the synthesis and functions of a porous catalytic framework based on conjugated micro- and mesoporous polymers with metalloporphyrin building blocks [3]. FeP-CMP was obtained using Suzuki polycondensation reaction (Fig. 5.3a). Based on its skeleton, FeP-CMP was attempted as a heterogeneous catalyst for the activation of molecular oxygen to convert sulfide to sulfoxide under ambient temperature and pressure (Fig. 5.3b and Table 5.1). FeP-CMP presents high efficiency indicated by high conversion (up to 99 %) and a large turnover number (TON = 97,320). In addition, it is widely applicable to various sulfides covering from aromatic to alkyl and cyclic substrates. Meanwhile it displays high selectivity to form corresponding sulfoxide under different reaction conditions.

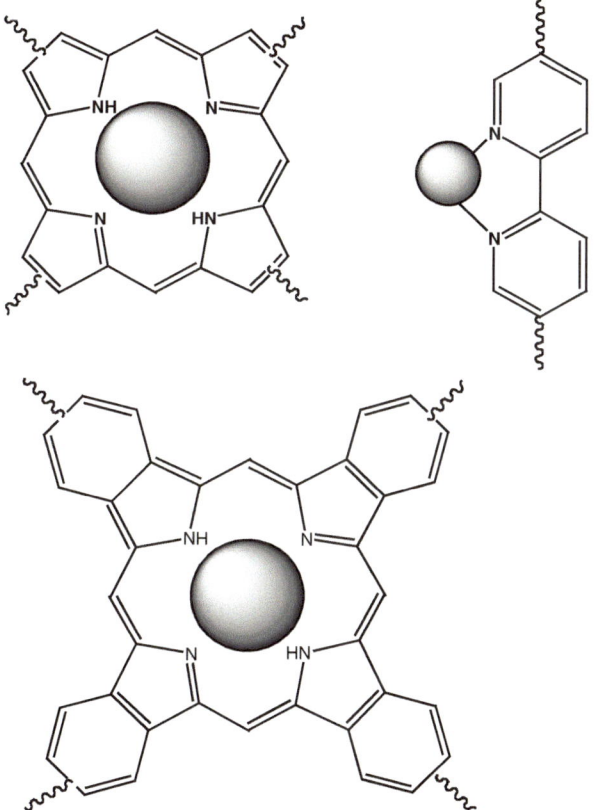

Fig. 5.2 Some typical building units with ability for coordinating metals

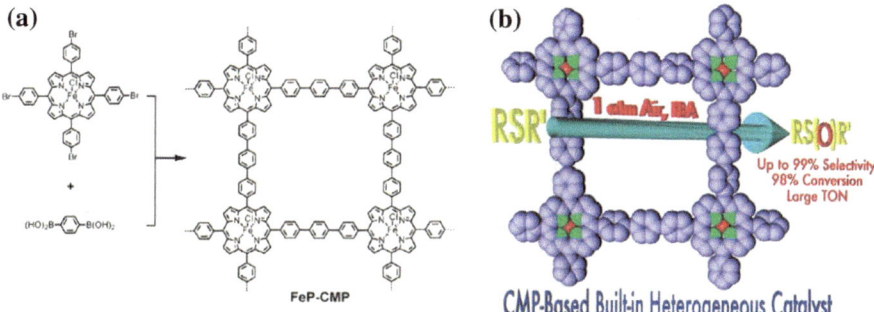

Fig. 5.3 **a** Schematic representation of the synthesis of nanoporous polymer with metalloporphyrin built-in skeleton (FeP-CMP). **b** Schematic representation of the transformation of sulfides to sulfoxides catalyzed by FeP-CMP. Reprinted with permission from Ref. [3]. Copyright 2011, American Chemical Society

Table 5.1 Catalytic oxidation of thioanisole by FeP-CMP. Reaction conditions: FeP-CMP (5.86×10^{-7} mol Fe), solvent (3 mL), room temperature, O_2 (1 atm)

Entry	Solvent	Time (h)	Conversion (%)	Selectivity		TON
				2	3	
1[a]	EtOAC	4.5	98	99	1	980
2[a]	THF	19	91	89	11	910
3[a]	CH_2Cl_2	22	98	99	1	980
4[a]	CH_3CN	24	97	98	2	970
5[a]	Toluene	1.5	97	98	2	970
6[b]	Toluene	48	0	–	–	–
7[c]	Toluene	48	0	–	–	–
8[d]	Toluene	60	38	>99	<1	380
9[e]	Toluene	48	33	>99	<1	330
10[f]	Toluene	48	70	>99	<1	700
11[g]	Toluene	4	94	99	1	940

Reprinted with permission from Ref. [3]. Copyright 2011, American Chemical Society
[a]Fe/thioanisole/IBA = 1/1,000/3,000 (molar ratio)
[b]In the absence of IBA
[c]Under N_2 (1 atm)
[d]In the absence of catalyst
[e]Fe/substrate/IBA = 1/1,000/1,000 (molar ratio)
[f]Fe/substrate/IBA = 1/1,000/2,000 (molar ratio)
[g]Under air (1 atm)

Cooper et al. reported two simple and versatile strategies for preparing metal–organic conjugated microporous polymers (MO-CMPs) (Figs. 5.4 and 5.5) [4]. The rhenium, rhodium, and iridium have been successfully incorporated into the CMP networks. The resulting networks are thermally robust and highly porous. A proof-of-concept study indicates that these metal-containing materials are promising for heterogeneous catalysis.

McKeown et al. present the first phthalocyanine-based synthesis of a PIM using a preformed chlorinated phthalocyanine (Fig. 5.6b), with the monomer 1 in a dibenzodioxane forming reaction to yield CoPc-PIM-B [5]. It displays permanent microporosity along with a moderate surface area with a value of 120 m^2 g^{-1}. Subsequently, CoPc-PIM-A with higher surface area (450–600 m^2 g^{-1} for Co) was obtained via the in situ metal-mediated synthesis of the desired cross-linked metallophthalocyanine from the rigid bis(phthalonitrile) precursor (Fig. 5.6a). In comparison with the low molar mass analog, the porous materials show considerably enhanced catalytic activity for many reactions, including H_2O_2 decomposition, cyclohexene oxidation. and hydroquinone oxidation.

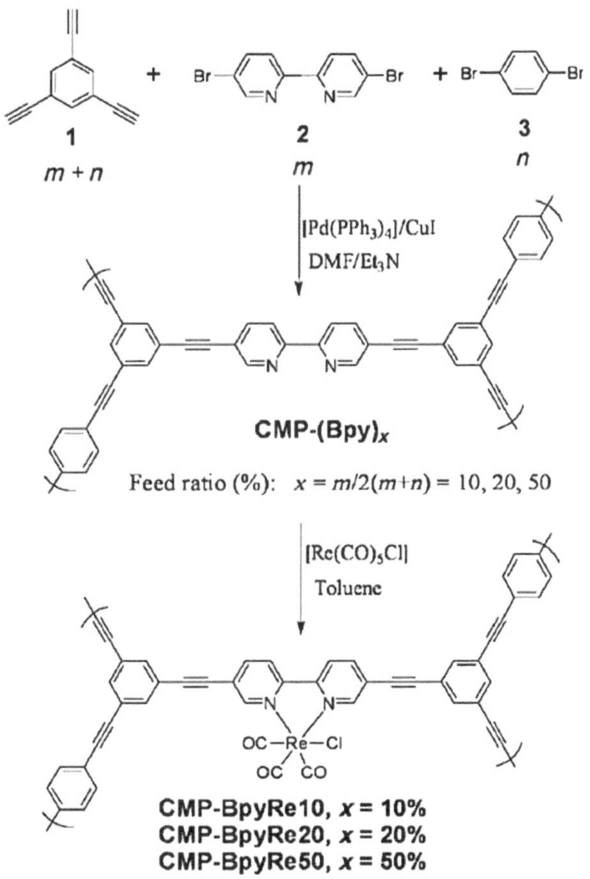

Fig. 5.4 Synthetic routes for the bipyridine-containing polymer precursors and the re-containing polymer networks. Reprinted with permission from Ref. [4]. Copyright 2011, Wiley-VCH

If the wall of POFs has special reactive groups that can "capture" the metals, the catalytic POFs could be synthesized by post-modified method. Nguyen et al. reported a catechol-functionalized porous organic polymer (POP) [6]. Post-modification has been successfully metallated with a Schrock-type TaV alkylidene and the resulting product remains thermally and structurally robust (Fig. 5.7). The POP-supported (catecholato)TaV alkyl sites remain accessible for small molecules and can undergo reactions to yield stable, monomeric complexes, which are quite different from those observed with the homogeneous analogs (Fig. 5.8). Such capabilities point toward opportunities to develop a new generation of well-defined supported monometallic catalysts whose environment can readily be understood along with their in situ reactivity.

Thomas et al. have developed a microporous polymer network, which was prepared from a weakly coordinating, anionic tecton [7]. The resulting material consists of immobilized, weakly coordinating tetraphenylborate ions. The anionic polymer network as a solid counterion for a catalytically active cationic species,

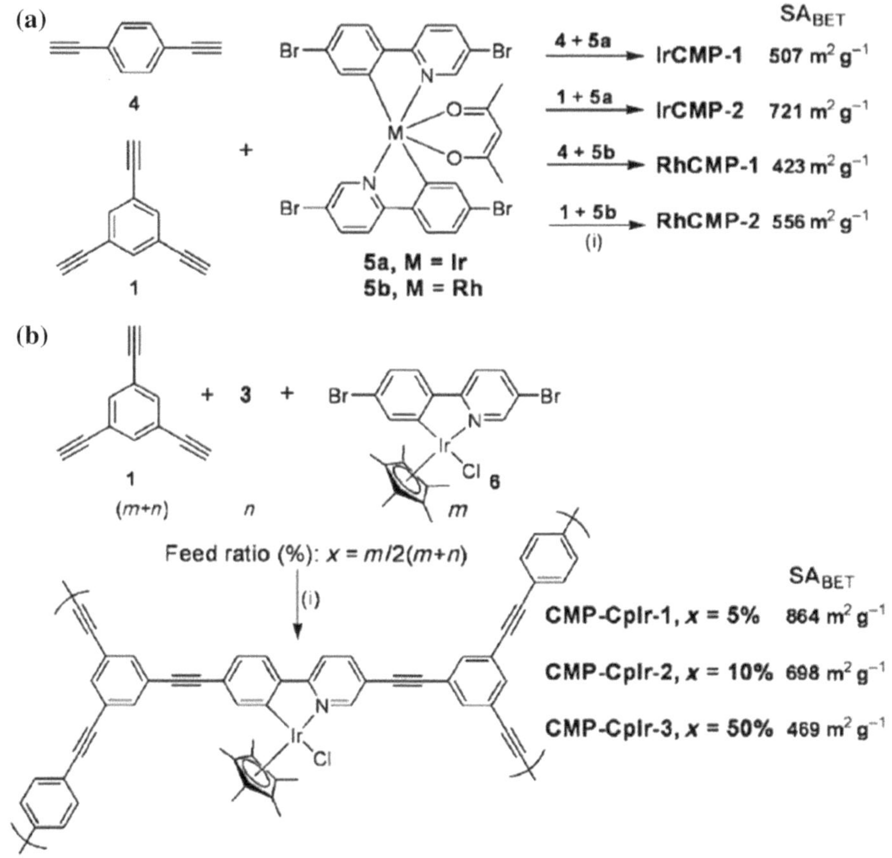

Fig. 5.5 **a** Direct synthetic routes for Ir- and Rh-containing MO-CMPs produced from cyclo-metalated Ir and Rh complexes, respectively; **b** MO-CMPs produced from Cp*-containing Ir monomer. *i* [Pd(PPh$_3$)$_4$], CuI, DMF, and Et$_3$N, 90 °C, 72 h. Reprinted with permission from Ref. [4]. Copyright 2011, Wiley-VCH

here [Mn(bpy)$_2$]$^{2+}$, was immobilized using a "ship-in-the-bottle" approach (Fig. 5.9). The resulting network is a promising catalyst for the aerobic oxidation of alkenes.

Nguyen and co-workers reported that an Al(porphyrin) functionalized with a large axial ligand was incorporated into a porous organic polymer using a cobalt-catalyzed acetylene trimerization strategy (Fig. 5.10) [8]. The presence of a large axial ligand on the Al(porphyrin) monomer can afford Al-PPOPs with larger pores that were more accessible to substrates in catalysis (Fig. 5.11), the best catalysts were obtained after supercritical CO$_2$ processing. In contrast to the conventional activation method of heating the samples under vacuum, supercritical CO$_2$ processing afforded POPs with much larger pores and total pore volumes, thus significantly enhancing substrate accessibility and catalytic rates.

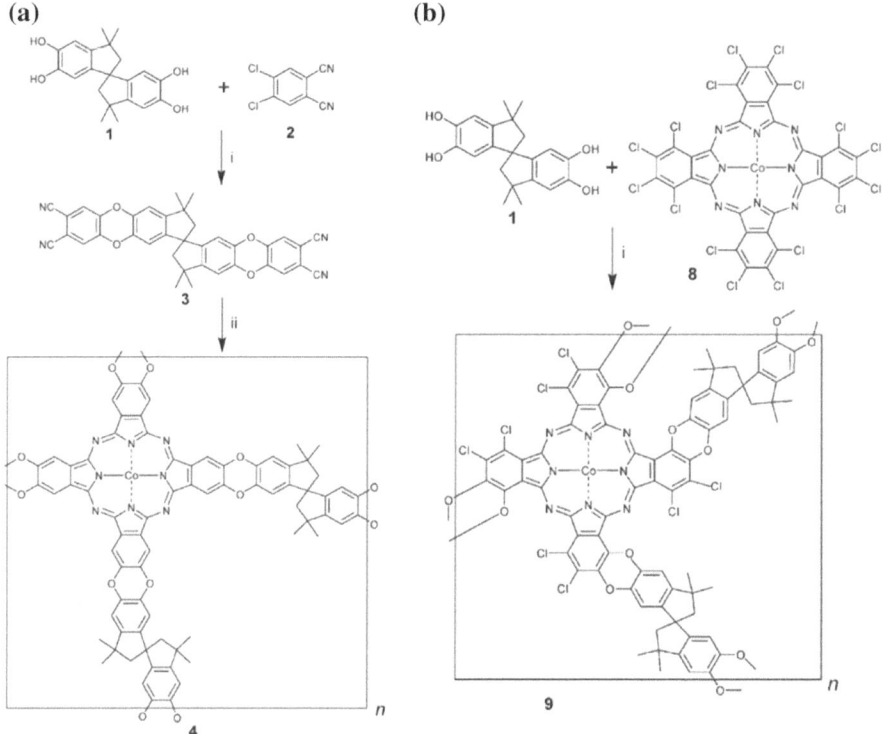

Fig. 5.6 **a** Preparation of spiro-linked cobalt phthalocyanine network polymer (CoPc-PIM-A) utilizing a phthalocyanine-forming reaction. Reagents and conditions: *i* K_2CO_3, DMF, 70 °C; *ii* $Co(CH_3COO)_2$, quinoline, 220 °C. **b** Preparation of spiro-linked cobalt phthalocyanine network polymer (CoPc-PIM-B) from preformed chlorinated phthalocyanine. Reagents and conditions: *i* K_2CO_3, NMP, 195 °C, 24 h. Reprinted with permission from Ref. [5]. Copyright 2008, Royal Society of Chemistry

5.2.2 POFs as Support of Catalyst

POFs have high surface area, high stability, and various structures; therefore, POFs could be considered as support of catalyst. One excellent example was illustrated by Wang et al. [9]. Crystalline COF material was first employed as catalyst support. COF-LZU1 is a two-dimensional 1imine-linked material (Fig. 5.12). The eclipsed layered-sheet structure makes the possibility of incorporation with metal ions. Via a simple post-treatment, a Pd(II)-containing COF, Pd/COF-LZU1, was accordingly synthesized, which showed excellent catalytic activity in catalyzing the Suzuki-Miyaura coupling reaction.

The structure of the POF material would determine that if it is suitable to be used as catalyst support. A study reported by Tan et al. shows a good example for illustrating the relationship between the structure and property [10]. The "knitting"

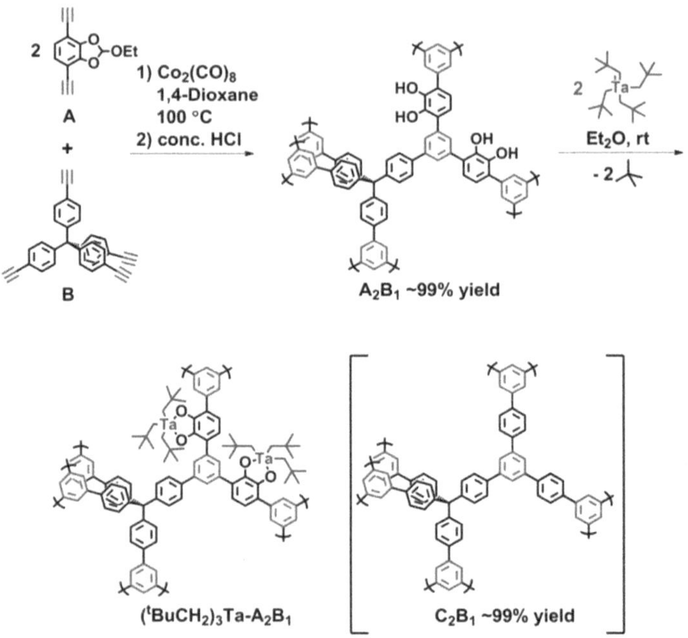

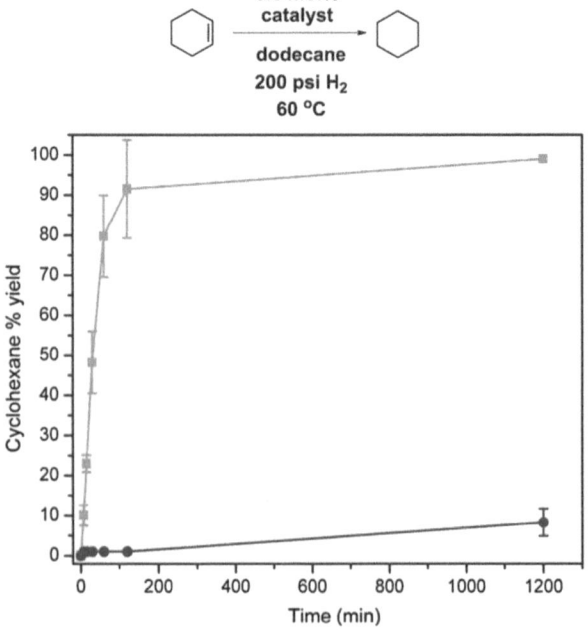

Fig. 5.7 Synthesis of catPOP A$_2$B$_1$ using a cobalt-catalyzed acetylene trimerization strategy and its subsequent metalation with (tBuCH$_2$)$_3$Ta]CHtBu. Shown on the *right-hand side* is the nonfunctionalized POP C$_2$B$_1$. The structures shown for the POPs are idealized representations. Reprinted with permission from Ref. [6]. Copyright 2013, Royal Society of Chemistry

Fig. 5.8 Yield of cyclohexane from the hydrogenation of cyclohexene (0.5 mol% Ta, 200 psi H$_2$, 60 °C in dodecane) with (tBuCH$_2$)$_3$Ta-A$_2$B1 (*red*) and (tBuCH$_2$)$_3$-Ta(3,6-tBu$_2$cat) (*blue*) at 0, 7.5, 15, 30, 60, 120, and 1,200 min. No other products were observed under these reaction conditions and times. Reprinted with permission from Ref. [6]. Copyright 2013, Royal Society of Chemistry

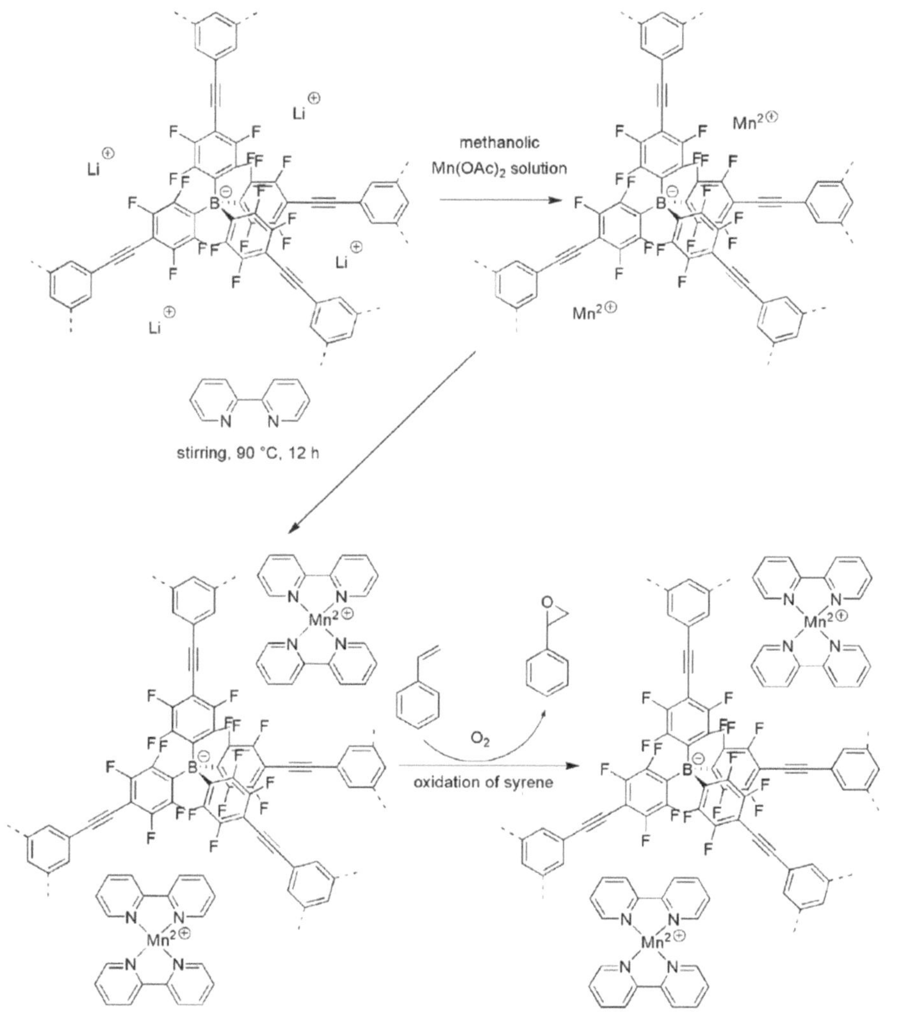

Fig. 5.9 Immobilization of [Mn(bpy)$_2$]$^{2+}$ in an ABN and its catalysis of styrene oxidation. Reprinted with permission from Ref. [7]. Copyright 2013, Wiley-VCH

strategy was utilized to prepare Pd(II) organometallic catalysts immobilized on the triphenylphosphine-functionalized KAPs(Ph-PPh$_3$) (Fig. 5.13). The porous structure provides the high dispersion of active Pd sites. In addition, heterogeneous porous structure also improved the diffusion of organic reactant molecules. Therefore, the resulted catalyst exhibited excellent activity and selectivity for the Suzuki–Miyaura cross-coupling reaction of aryl chlorides even under mild conditions and aqueous reaction media. KAPs(Ph-PPh$_3$)-Pd could satisfy the demand on an industrial scale owing to various advantages, including nontoxic aqueous media, mild reaction conditions (80 °C), stability of the catalyst, and the facile

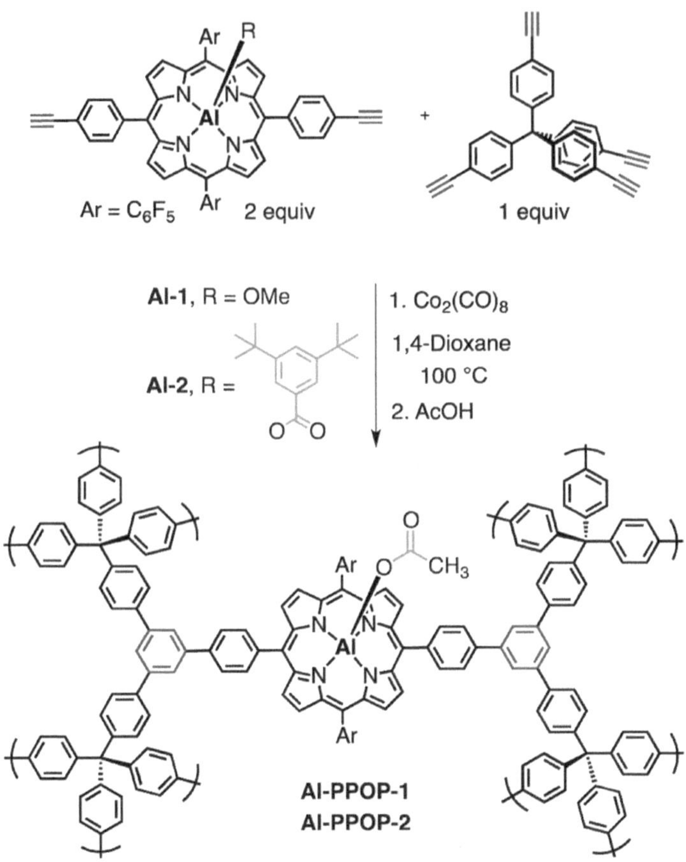

Fig. 5.10 Synthesis of Al-PPOPs via a cobalt-catalyzed acetylene trimerization. Reprinted with permission from Ref. [8]. Copyright 2013, American Chemical Society

synthesis approach (Table 5.2). Furthermore, this work also highlights that the microporous polymers not only play the roles as support materials, but also protect the catalyst and positively affect the catalytic activity.

5.3 POFs Without Metal Site for Catalysis

In comparison with metal-containing catalytic POFs, the design and synthesis of metal-free catalytic POFs is still a challenge. One effective strategy is based on the successful examples of small organic molecules for catalytic reactions. Herein, we will introduce some representative metal-free catalytic POFs.

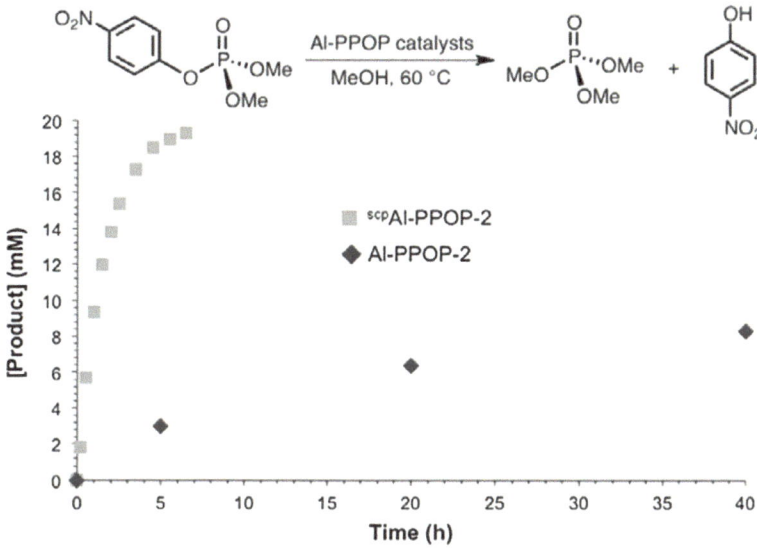

Fig. 5.11 Reaction profiles for the methanolysis of PNPDPP in the presence of 4 mol% of Al-PPOP-2 (*blue diamonds*) and scpAl-PPOP-2 (*red squares*). Reprinted with permission from Ref. [8]. Copyright 2013, American Chemical Society

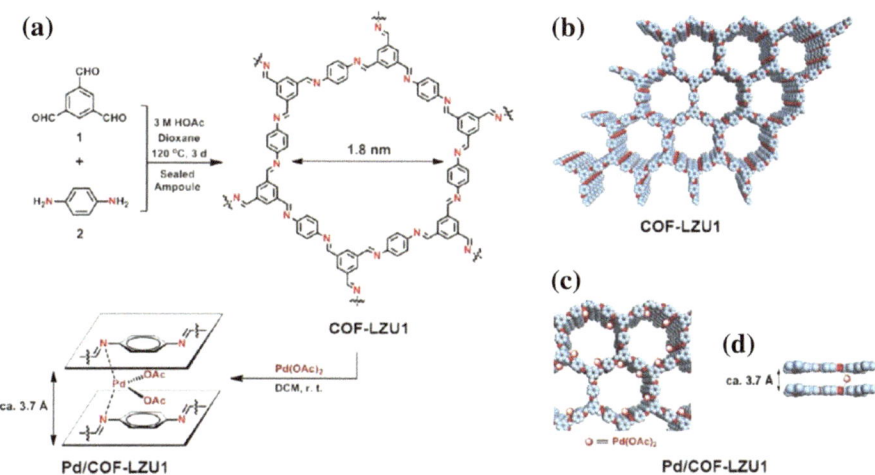

Fig. 5.12 Construction of COF-LZU1 and Pd/COF-LZU1. Schematic representation for the synthesis of COF-LZU1 and Pd/COF-LZU1 materials (**a**). Proposed structures of COF-LZU1 (**b**) and Pd/COF-LZU1 (**c, d**) possessing regular microporous channels (diameter of ∼1.8 nm), simulated with a 2D eclipsed layered-sheet arrangement. C *blue*, N *red*, and *brown spheres* represent the incorporated Pd(OAc)₂. Reprinted with permission from Ref. [9]. Copyright 2011, American Chemical Society

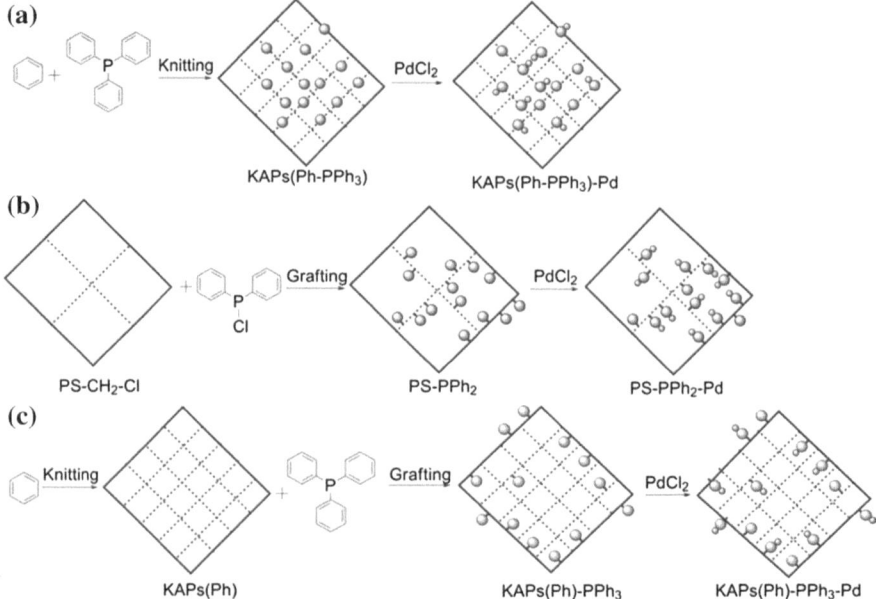

Fig. 5.13 Synthetic route of: **a** KAPs(Ph-PPh₃)-Pd, **b** PS-PPh₂-Pd and, **c** KAPs(Ph)-PPh₃-Pd. *Big balls* represent PPh₃ or PPh₂ functional groups. *Small balls* represent PdCl₂. *Solid line* represents boundary of material particles. *Dashed lines* represent inner network of material particles. Reprinted with permission from Ref. [10]. Copyright 2012, Wiley-VCH

5.3.1 Jorgensen–Hayashi Catalyst

Wang et al. reported one of the most versatile organocatalysts [11]. Via a bottom-up strategy, the Jorgensen–Hayashi catalyst has been successfully designed and embedded into a nanoporous polymer JH-CPP (Fig. 5.14). JH-CPP has high BET surface area with a value of 881 $m^2 \, g^{-1}$, wide openings, and interconnected pores. Increasing the accessibility of catalytic sites and facilitating the mass transport process ensured JH-PP possesses excellent catalytic activity. JH-CPP catalyst was utilized for catalyzing asymmetric Michael addition of aldehydes to nitroalkenes. It gave desired products in good to high yield (67–99 %), high enantioselectivity (93–99 % ee), and high diastereoselectivity (d.r. of 74:26–97:3) (Table 5.3). In addition, the JH-CPP catalyst showed high stability, indicated by the recycling experiment. JH-CPP could be reused at least for four times without the loss of enantioselectivity (97–99 % ee) and diastereoselectivity (d.r. 92:8–88:12).

Table 5.2   Suzuki–Miyaura coupling reaction catalyzed by KAPs(Ph-PPh$_3$)-Pd[a]

$$R^1-\!\!\!\left\langle\!\!\bigcirc\!\!\right\rangle\!\!-X \; + \; (HO)_2B-\!\!\left\langle\!\!\bigcirc\!\!\right\rangle \quad\xrightarrow[\substack{K_3PO_4\cdot3H_2O\\H_2O/EtOH}]{0.6\ mol\%\ Pd,\ 80\ ^\circ C}\quad R^1-\!\!\left\langle\!\!\bigcirc\!\!\right\rangle\!\!-\!\!\left\langle\!\!\bigcirc\!\!\right\rangle$$

Entry	Substrate 1	Substrate 2	Time (h)	Yield[b] (%)
1	Ph–Cl	(HO)$_2$B–Ph	3	99
2	H$_3$COC–C$_6$H$_4$–Cl	(HO)$_2$B–Ph	16	84
3	O$_2$N–C$_6$H$_4$–Cl	(HO)$_2$B–Ph	4	97
4	2-NO$_2$-C$_6$H$_4$–Cl	(HO)$_2$B–Ph	6	84
5	3-O$_2$N-C$_6$H$_4$–Cl	(HO)$_2$B–Ph	4	88
6	2-CN-C$_6$H$_4$–Cl	(HO)$_2$B–Ph	10	99
7	CH$_3$–C$_6$H$_4$–Cl	(HO)$_2$B–Ph	4	91
8	2-CH$_3$-C$_6$H$_4$–Cl	(HO)$_2$B–Ph	4	87
9	H$_3$CO–C$_6$H$_4$–Cl	(HO)$_2$B–Ph	12	65
10	Ph–Br	(HO)$_2$B–Ph	1	97
11	Ph–I	(HO)$_2$B–Ph	1	98
12	Ph–CH$_2$Cl	(HO)$_2$B–Ph	2	95[c]
13	Ph–Cl	(HO)$_2$B–C$_6$H$_4$–CH$_3$	4	97
14	Ph–Cl	(HO)$_2$B–C$_6$H$_4$–CF$_3$	6	73

Reprinted with permission from Ref. [10]. Copyright 2012, Wiley-VCH
[a]0.6 mol% Pd, 0.5 mmol of aryl halide, 0.75 mmol of arylboronic acid, 1.5 mmol of K$_3$PO$_4 \cdot$ 3H$_2$O, 2 mL of EtOH:H$_2$O (3:2 v/v), 80 °C, under a N$_2$ atmosphere
[b]Isolated yield
[c]0.5 mmol of benzyl chloride was used

5.3.2 Photocatalytic POFs

Cooper et al. have shown the synthesis of microporous organic networks (MONs) by Sonogashira coupling reaction using multialkyne connectors and multih-aloarene monomers [12]. Son and Lee et al. have continued to develop functional MONs [13]. They present a tandem synthetic strategy for successful construction of POFs with active species [14]. The generation of benzodifuran species could

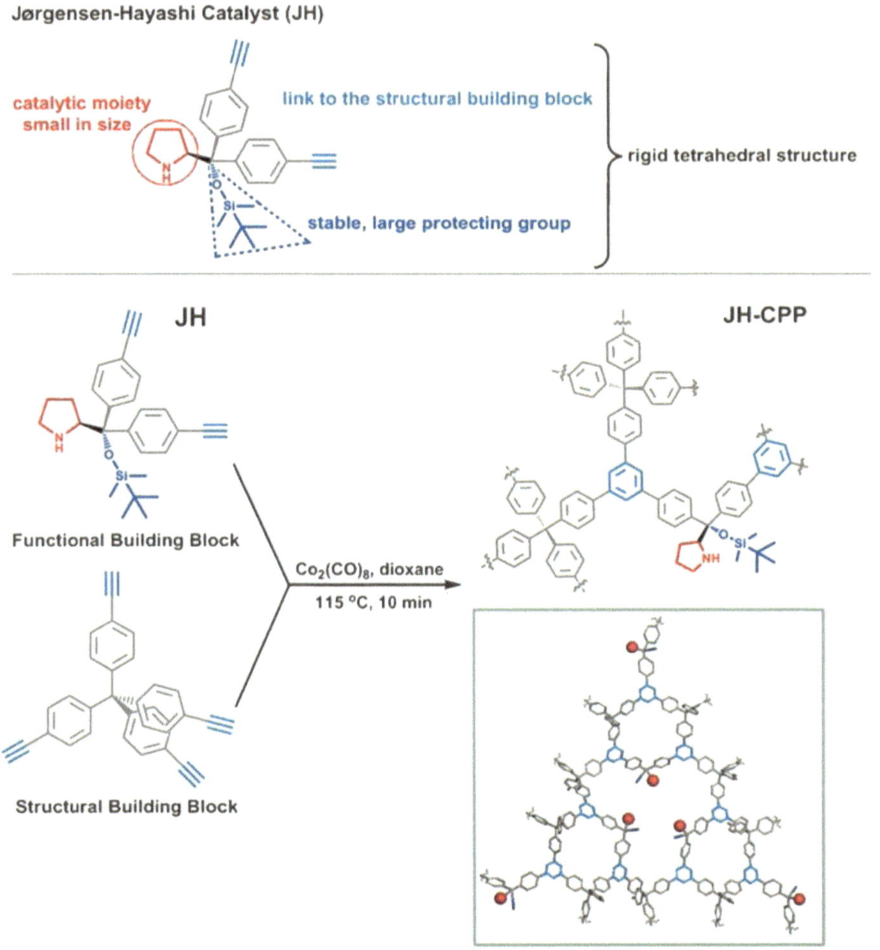

Fig. 5.14 Rational design and synthesis of a chiral porous polymer (*JH-CPP*) embedded with Jørgensen–Hayashi catalyst (*JH*) by the bottom-up strategy. The structural uniqueness of the JH catalyst as the functional building block is depicted (*top*). Through the $Co_2(CO)_8$-mediated trimerization, the JH-CPP polymer could befacilely synthesized from JH and the structural building block. The simulated local structure of JH-CPP (*bottom*) shows the pore formation and the distribution of catalytic moieties (*red*). Reprinted with permission from Ref. [11]. Copyright 2012, Wiley-VCH

be induced in a tandem manner during the formation of the MON (Fig. 5.15a). Catalyzed by the resulting benzodifuran-containing BDF-MON, the oxidative conversion of primary amines into imines was achieved under the photocatalytic activities. Employing this delicate combination of network-forming reactions for POFs, more diverse tandem synthetic approaches for functional POFs can be designed.

Table 5.3 Screening the solvents for the asymmetric Michael addition reaction catalyzed by JH-CPP[a]

Entry	Solvent	t (h)	Yield (%)[b]	Syn/anti[c]	ee (%)[d]
1	n-hexane	16	26	89:11	93
2	CH_2Cl_2	20	24	90:10	99
3	MeOH	3	75	89:11	98
4	H_2O	12	89	85:15	98
5	EtOH	10	96	84:16	94
6	EtOH/H_2O (1:1)	2	96	92:8	99
7[e]	EtOH/H_2O (1:1)	24	64	78:22	97

Reprinted with permission from Ref. [11]. Copyright 2012, Wiley-VCH
[a]General condition: (E)-4-chloro-b-nitrostyrene (0.1 mmol), propanal (1.0 mmol), JH-CPP (0.01 mmol), solvent (0.5 mL), benzoic acid (0.1 mmol), RT
[b]Isolated yield after silica gel column chromatography
[c]Determined by 1H NMR spectroscopy
[d]Determined by chiral HPLC (Daicel chiral AD-H column)
[e]No benzoic acid

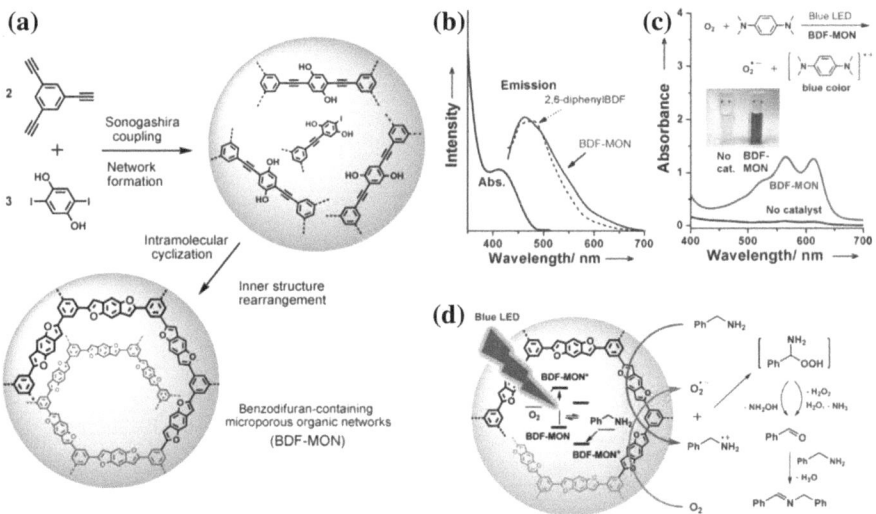

Fig. 5.15 a Strategy for the synthesis of microporous organic networks; **b** UV/Vis absorption and emission spectra (exc. 410 nm) of BDF-MON and emission spectrum of 2,6-diphenylbenzodifuran. The emission from medium (CH_3CN) was calibrated; **c** UV/Vis absorption spectra and photograph of the cationic radical species of 1,4-bis(dimethylamino) benzene (100 mm in 3 mL CH_3CN) generated by BDF-MON (21 mg) and irradiation with a blue LED for 1 h in the presence of oxygen; **d** a suggested photocatalytic process of oxidative conversion of benzylamine into imine by BDF-MON under irradiation with a blue LED. Reprinted with permission from Ref. [14]. Copyright 2013, Wiley-VCH

5.3.3 Bifunctional Solid Catalyst

The unique feature of enzymes is that they could catalyze organic molecules via multistep reactions. Cascade reactions could perform a consecutive series of reactions to proceed in a concurrent fashion without isolation of intermediates and self-quenching of catalysts. Therefore, the development of bi/multifunctional solid catalysts for heterogeneous cascade catalysis has attracted great attention.

PAF-1 has high surface area, large pore size, and high stability. Therefore PAF-1 is a suitable starting material for post-modification. Ma et al. illustrated dual functionalization of PAF-1 [15]. That is in the context of incorporating two antagonistic sites of strong acid and strong base into PAF-1 via stepwise post-synthetic modification (Fig. 5.16). Compared to the counterparts of mesoporous silica and MOFs, the resulting bifunctionalized PAF-1 shows an excellent catalytic performance in a series of cascade reactions (Table 5.4). In addition, bifunctionalized PAF-1 demonstrates superior chemical stability. Thereby, the door for dual functionalization of PAFs as a new platform for heterogeneous cascade catalysis is opened.

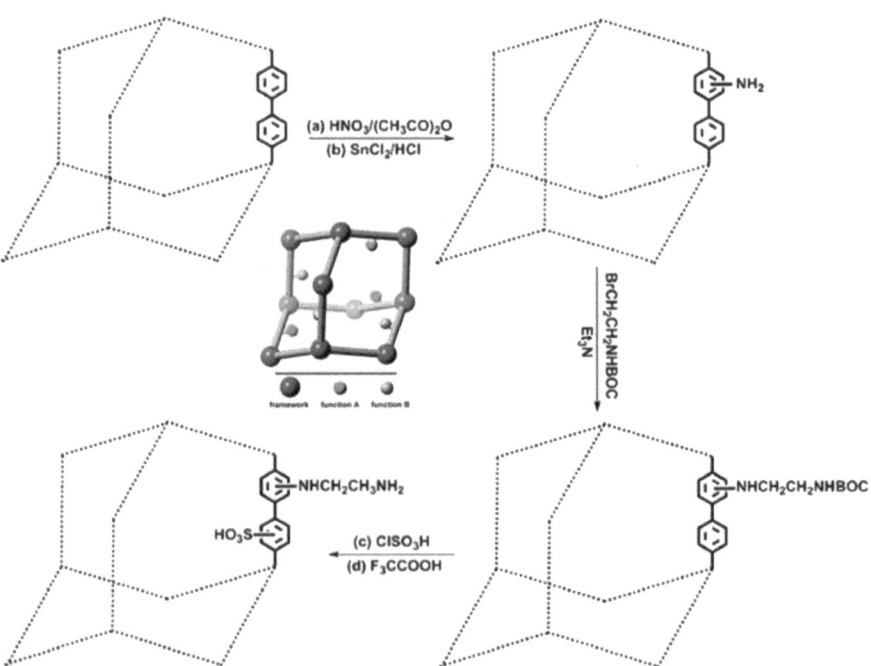

Fig. 5.16 Illustration of stepwise post-synthetic modification of PAF-1 to graft the two antagonistic functional groups of a strong acid and a strong base to afford PAF-1-NHCH$_2$CH$_2$NH$_2$-SO$_3$H. Reprinted with permission from Ref. [15]. Copyright 2014, Royal Society of Chemistry

Table 5.4 One pot deacetalization–Henry reaction[a]

Entry	Catalyst	Conv. of 1 (%)	Yield of 2 (%)	Yield of 3 (%)
1	PAF-1-NHCH$_2$CH$_2$NH$_2$-SO$_3$H	100	2	97.2
2	PAF-1-SO$_3$H	100	100	0
3	PAF-1-CH$_2$NH$_2$	Trace	Trace	Trace
4	PAF-1-NH$_2$-SO$_3$H	100	Trace	Trace
5	Ethylamine + p-toluene sulfonic acid	Trace	Trace	Trace
6	PAF-1-NHCH$_2$CH$_2$NH$_2$-SO$_3$H + ethylamine	Trace	Trace	Trace
7	PAF-1-NHCH$_2$CH$_2$NH$_2$-SO$_3$H + p-toluene sulfonic acid	100	100	Trace
8[b]	PAF-1-NHCH$_2$CH$_2$NH$_2$-SO$_3$H	100	5	94.7

[a]Reaction conditions: benzaldehyde dimethyl acetal (1.0 mmol), CH$_3$NO$_2$ (5.0 mL), 90 °C, 24 h
[b]Treatment with NaOH (2 M) and HCl (2 M)

5.3.4 Shape-Selective Catalyst

It is clear that POFs have presented excellent performance for catalyzing different reactions according to the above introduction. However, their amorphous textures result in their irregular pores. Therefore it is a challenge to employ them for catalyst using the selection of pore size. Yan reported the design and synthesis of two new 3D microporous base-functionalized COFs, termed BF-COF-1 and BF-COF-2 [16]. PXRD results show BF-COF-1 and BF-COF-2 are highly crystalline. Based on the structural analysis, BF-COF-1 has microporous cavities with a diameter of 7.8 Å and rectangular windows with a size of 7.8 × 11.3 Å^2, whereas BF-COF-2 exhibits microporous cavities of 7.7 Å with rectangular windows of 7.7 × 10.5 Å^2 (Fig. 5.17). Based on the N$_2$ sorption isotherms, the pore-size distributions of the BF-COFs were calculated using nonlocal density functional theory (NLDFT) (Fig. 5.18). They present a narrow pore distribution (8.3 Å for BF-COF-1 and 8.1 Å for BF-COF-2), which match well with the pore size predicted from the crystal structures. Both BF-COFs are employed as catalysts in the Knoevenagel condensation reaction, showing remarkable conversion (96 % for BF-COF-1 and 98 % for BF-COF-2) (Fig. 5.19). It is notable that the catalyst presents high size selectivity indicated by the different selected molecules, and good recyclability. This study suggests that porous functionalized 3D COFs could be a promising new class of

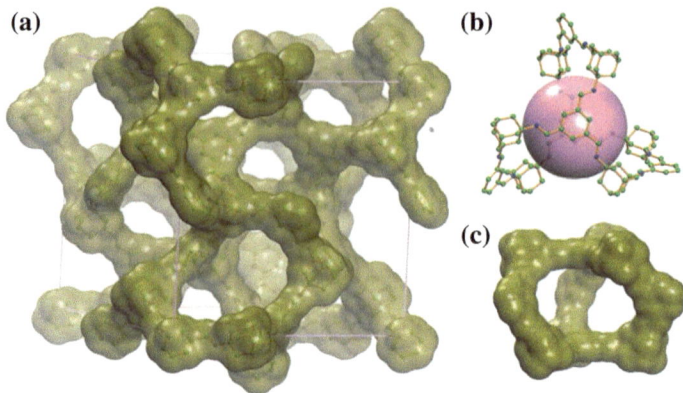

Fig. 5.17 Extended structure of BF-COF-1. **a** Atomic connectivity in BF-COF-1 with atom surfaces in *olive green*. Carbon and nitrogen are represented as *green* and *blue spheres*, respectively; **b** a microporous cavity (*pink sphere*) with a diameter of 7.8 Å in BF-COF-1; **c** Rectangular windows with a diameter of 7.8×11.3 Å^2 in BF-COF-1. Reprinted with permission from Ref. [16]. Copyright 2014, Wiley-VCH

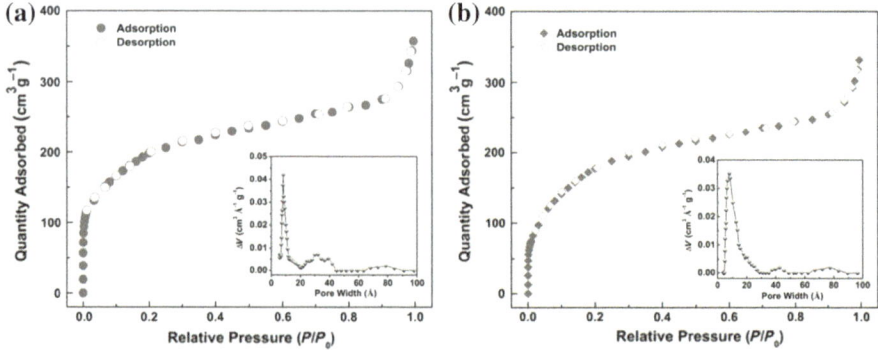

Fig. 5.18 **a** Nitrogen-gas adsorption (*filled symbols*) and desorption isotherms (*open symbols*) measured at 77 K for **a** BF-COF-1 and **b** BF-COF-2. *Inset* Pore-size distribution of the BF-COFs, as calculated by fitting NLDFT models to the adsorption data. Reprinted with permission from Ref. [16]. Copyright 2014, Wiley-VCH

shape-selective catalysts. The successful synthesis of 3D porous functionalized COFs would greatly facilitate the development of COFs as promising selective catalysts.

Owing to the various advantages of POF materials, including diversiform porous structures, high stability, and adjustable skeletons, POFs display excellent

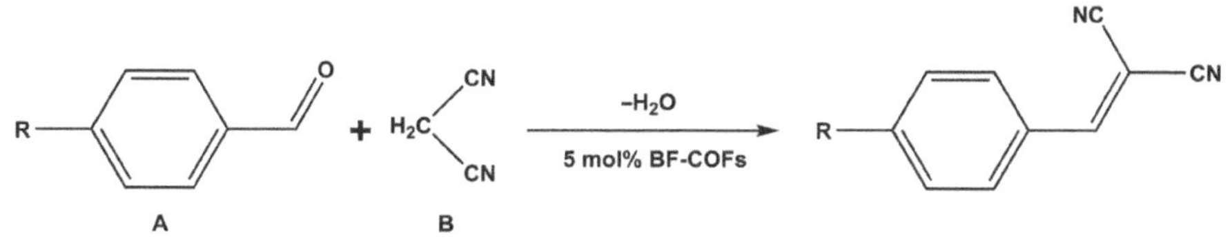

Reaction	A	B	Catalyst	Conversion (%)
I	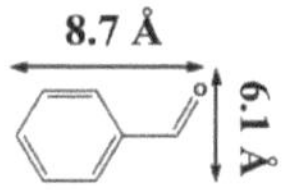	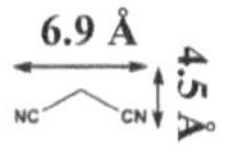	BF-COF-1	96
			BF-COF-2	98
II	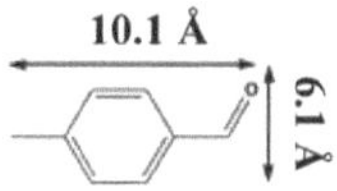	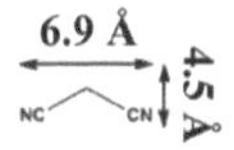	BF-COF-1	4
			BF-COF-2	3
III	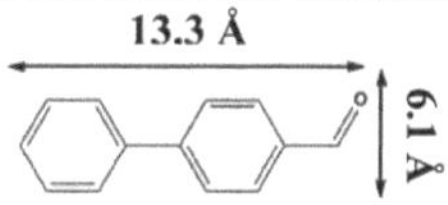	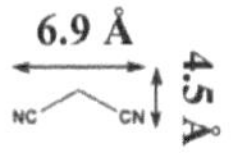	BF-COF-1	3
			BF-COF-2	4
IV	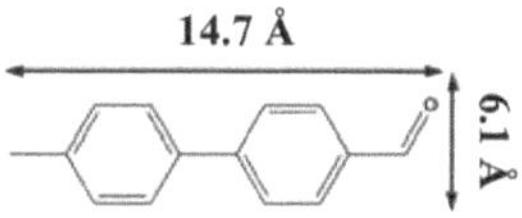	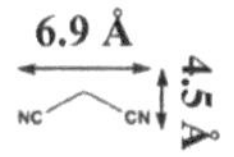	BF-COF-1	2
			BF-COF-2	2

Fig. 5.19 Catalytic activity of BF-COFs in the Knoevenagel condensation reaction. Reprinted with permission from Ref. [16]. Copyright 2014, Wiley-VCH

performance for catalyst. The POFs could be considered as promising materials for the applications in catalytic field.

References

1. Kaur P, Hupp J, Nguyen S et al (2011) Porous organic polymers in catalysis: opportunities and challenges. ACS Catal 1:819–835
2. Zhang Y, Riduan S (2012) Functional porous organic polymers for heterogeneous catalysis. Chem Soc Rev 41:2083–2094
3. Chen L, Yang Y, Jiang D et al (2010) CMPs as scaffolds for constructing porous catalytic frameworks: a built-in heterogeneous catalyst with high activity and selectivity based on nanoporous metalloporphyrin polymers. J Am Chem Soc 132:9138–9143

4. Jiang J, Wang C, Laybourn A et al (2011) Metal–organic conjugated microporous polymers. Angew Chem Int Ed 50:1072–1075
5. Mackintosh H, Budd P, McKeown N et al (2008) Catalysis by microporous phthalocyanine and porphyrin network polymers. J Mater Chem 18:573–578
6. Tanabe K, Siladke N, Broderick E et al (2013) Stabilizing unstable species through single-site isolation: a catalytically active TaV trialkyl in a porous organic polymer. Chem Sci 4:2483–2489
7. Fischer S, Schmidt J, Strauch P et al (2013) An anionic microporous polymer network prepared by the polymerization of weakly coordinating anions. Angew Chem Int Ed 52:12174–12178
8. Totten R, Kim Y, Weston M et al (2013) Enhanced catalytic activity through the tuning of micropore environment and supercritical CO_2 processing: Al (Porphyrin)-based porous organic polymers for the degradation of a nerve agent simulant. J Am Chem Soc 135:11720–11723
9. Ding S, Gao J, Wang Q et al (2011) Construction of covalent organic framework for catalysis: Pd/COF-LZU1 in Suzuki–Miyaura coupling reaction. J Am Chem Soc 133:19816–19822
10. Li B, Guan Z, Wang W et al (2012) Highly dispersed Pd catalyst locked in knitting aryl network polymers for Suzuki–Miyaura coupling reactions of aryl chlorides in aqueous media. Adv Mater 24:3390–3395
11. Wang C, Zhang Z, Yue T et al (2012) "Bottom-up" embedding of the jørgensen-hayashi catalyst into a chiral porous polymer for highly efficient heterogeneous asymmetric organocatalysis. Chem Eur J 18:6718–6723
12. Jiang J, Su F, Trewin A et al (2007) Conjugated microporous poly(aryleneethynylene) networks. Angew Chem Int Ed 46:8574–8578
13. Kang N, Park J, Choi J et al (2012) Nanoparticulate iron oxide tubes from microporous organic nanotubes as stable anode materials for lithium ion batteries. Angew Chem Int Ed 51:6626–6630
14. Kang N, Park J, Ko K et al (2013) Tandem synthesis of photoactive benzodifuran moieties in the formation of microporous organic networks. Angew Chem Int Ed 52:6228–6232
15. Zhang Y, Li B, Ma S et al (2014) Dual functionalization of porous aromatic frameworks as a new platform for heterogeneous cascade catalysis. Chem Commun 50:8507–8510
16. Fang Q, Gu S, Zheng J et al (2014) 3D microporous base-functionalized covalent organic frameworks for size-selective catalysis. Angew Chem 126:2922–2926

Chapter 6
Other Applications of Porous Organic Frameworks

Abstract It has been documented that POFs present excellent performance in the fields of gas storage, gas separation, and catalyst, etc. Because of the facile constructing principles, POFs could be prepared by controlling the original building units to satisfy special demands. The versatile structural skeletons of POFs lead to other glamorous properties. In this chapter, we introduce some characters of POFs, including light-harvesting, electroactive property, and applications in solar cells, antibacterial, iodine capture, etc.

Keywords Host–guest chemistry · Light-harvesting · Semiconduction · Photo-conduction · Solar cell · Antibacterial · Iodine capture · Oxygen reduction reaction

6.1 Introduction

As we know, the structure of materials determines their corresponding properties. As mentioned above, it has been documented that POFs present excellent performance in the fields of gas storage, gas separation, catalyst, etc. [1, 2]. POFs are composed of organic building blocks via polymerization reactions. The facile construction principles allow us to design and synthesize controllable structure of POFs with special functions. The versatile functions could be imparted through the selection of building units. Given diverse structural skeletons, porosities, and controlled compositions, POFs also display other glamorous properties. In this chapter, we will introduce some representative POFs, indicating the relationship between structure and function.

© The Author(s) 2015
G. Zhu and H. Ren, *Porous Organic Frameworks*, SpringerBriefs in Green Chemistry for Sustainability, DOI 10.1007/978-3-662-45456-5_6

6.2 Light-Harvesting

Conjugated microporous polymers (CMPs) have attracted much attention as they are able to elaborate integration of π-electronic components to the conjugated framework system and remain permanent porous structures at the same time [3–5]. Thus the energy donating CMPs with highly dense π-electronic components could be employed as antennae for the collection of photons. Additionally, energy-accepting counterparts could be spatially confined by the inherent pores of CMPs. Therefore, an unprecedented donor–acceptor system could be created for energy transduction mediated by CMP networks.

In 2010 Jiang et al. described the synthesis of light-harvesting CMP with rapid and highly efficient flow of light energy [6]. The polyphenylene-based CMP (PP-CMP) was synthesized through Suzuki coupling reaction (Fig. 6.1a). In comparison with the linear polyphenylene, PP-CMP is an inherent porous structure which consists of conjugated three-dimensional polyphenylene scaffolds. It possesses large surface area which is up to 1,083 m^2 g^{-1}. It is a blue photoluminescence emitting compound, which is capable of excitation energy migration over the framework. Interestingly, the porous structure of PP-CMP results in a

Fig. 6.1 a Structure representation of PP-CMP and coumarin 6 and **b** schematic representation of energy funneling from PP-CMP to spatially confined coumarin 6. Reprinted with permission from Ref. [6]. Copyright 2010, American Chemical Society

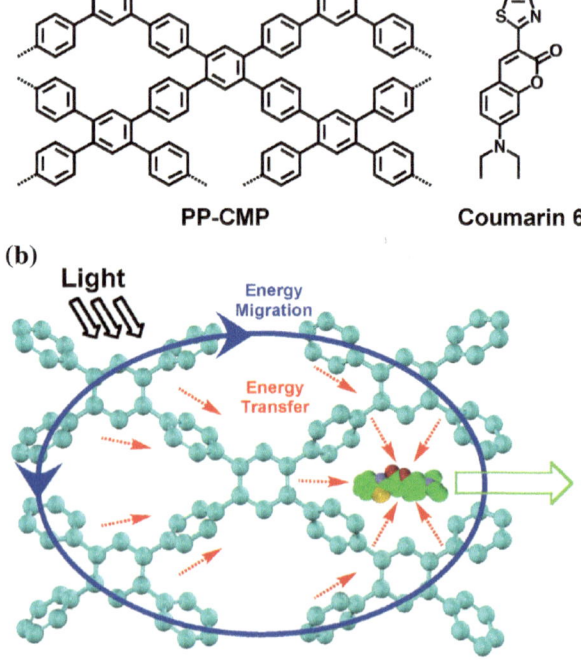

PP-CMP   **Coumarin 6**

Light Harvesting with a Porous Polymer Framework

spatial confinement of energy-accepting coumarin 6 molecules in their channels (Fig. 6.1b). As a result, a light-harvesting system with designable donor–acceptor compositions is obtained. Brilliant green emission from coumarin 6 was achieved by excitation of the PP-CMP skeleton. Compared with direct excitation of coumarin 6, the excitation from PP-CMP can reach much higher intensity which is up to 21 times. At the same time, the fluorescence of PP-CMP is almost quenched owing to the energy transfer from PP-CMP framework to coumarin 6. It suggests that PP-CMP framework plays a role for light-harvesting, which originates from the conjugated porous structure.

6.3 Electronic and Photonic-Electronic Properties

Due to their unique and attractive properties, organic semiconductor materials are extensively investigated. By now, huge amounts of organic semiconductor materials can be efficiently produced and processed on a large scale [7–9]. However, the low stability and efficiencies of the devices are still key issues to be addressed. Thereinto, the slow charge carrier mobility and fast recombination always result in low efficiencies of the materials. One of the possible reasons should be the inefficient stacking of the conducting polymers or disordered donor–acceptor interfaces. Therefore, conducting materials with total controlled nanoscale structure and orientation would be desirable for a higher conversion capability. The facile strategy of designing the functional porous organic framework materials on a molecular basis permits to adjust the energy gap and the light absorbance which might provide a good future of organic semiconductor.

6.3.1 Theoretical Insights

One of the most important advantages for COFs is that they are ordered porous structures with high surface area. Heine group investigated the structure and energy of the reported 2D covalent organic frameworks based on the density functional tight-binding (DFTB) theory [10]. According to their simulation, the stacking shapes could be classified into four different types, including AA, AB, serrated, and inclined (Fig. 6.2). The structure of the synthesized product was examined by comparison between theoretical simulation and PXRD pattern. However, the theoretically simulated diffraction patterns of all the different stacking types possess similar peak positions. Thus the structures of experimental materials cannot be well distinguished. Then, stacking energies which play a pivotal role were calculated. As all 2D covalent organic frameworks are semiconductors, their similar band gaps between 1.7 and 4.0 eV depend on the amount of aromatics groups and other factors.

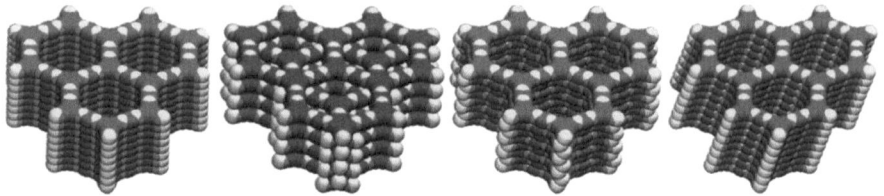

Fig. 6.2 Schematic of the stacking types: AA, AB, serrated, and inclined. Reprinted from Ref. [10]

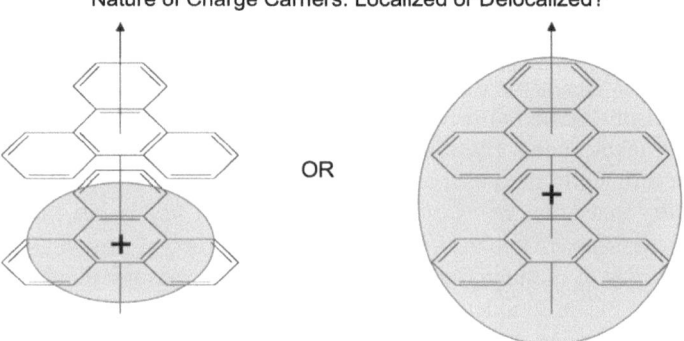

Fig. 6.3 Nature of charge carriers: localized or delocalized. Reprinted with permission from Ref. [11]. Copyright 2011, American Chemical Society

In 2011, Siebbeles group introduced a new modular method to measure the aromatic delocalizing electronic structure in triphenylene-based COFs [11]. The results show that the electronic coupling of interlayers depends on the molecular distance and the twist angle (Fig. 6.3). The superior electronic interaction will happen in the ecliptically stacked COF layers. In addition, delocalized charges could move by the temperature-dependent mechanism. As expected, COF materials show very high charge carrier mobilities. Actual mobility studies in diverse conditions demonstrate the charge transport mechanism. Such a vast number of calculations and experimental investigations on optoelectronic properties laid the foundation for organic solar cells and optoelectronic applications.

6.3.2 Semiconduction

As reported, the shape of 2D COFs is a layered structure with hexagonal or tetragonal sheets [12]. The primary driving force is driven by the out-of-plane p interactions. 2D COFs contains plenty of electronic couplings between the p-orbitals of the stacking layers. It is favorable to transport the charge carriers and photoexcited

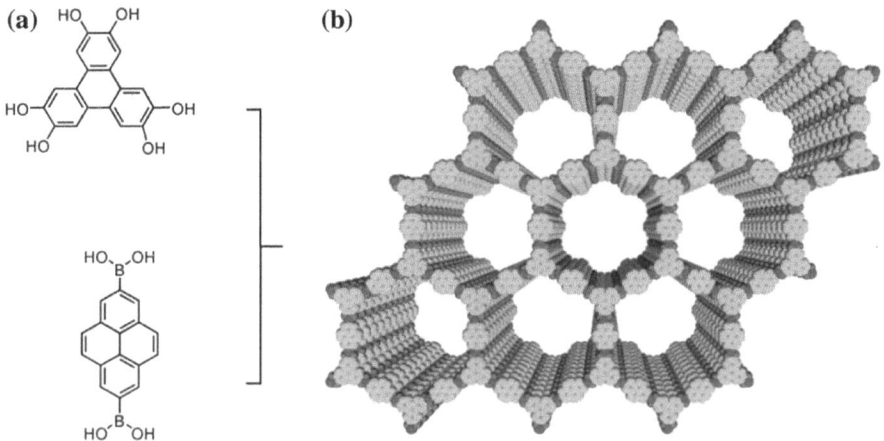

Fig. 6.4   a The monomers for the synthesis of TP-COF; **b** schematic representation of TP-COF (structure is based on quantum calculation and crystal lattice parameters; B *purple*, O *red*, tri-phenylene *green*, pyrene *blue*, H atoms are omitted for clarity). Reprinted with permission from Ref. [13]. Copyright 2008, Wiley-VCH

states (excitons). Therefore, 2D COFs could be considered as a new platform to molecularly design semiconducting and photoconducting materials.

In this research area, Jiang et al. have made a large contribution. In 2008, they demonstrated the use of COFs for the construction of electronic and optoelectronic materials [13]. It is notable that TP-COF is the first material for this novel property. The p-electronic 2D TP-COF consisting of interlocking hexagons is synthesized using HTTP molecules and pyrene-2,7-diboronic acid (PDBA) monomers (Fig. 6.4). Based on the XRD pattern and simulations, structural determinations show that the planar sheet is layered. TP-COF can harvest a wide range of photons from the ultraviolet to the visible regions. Therefore, fluorescence microscopy shows that TP-COF has an intense blue luminescence owing to the excitation of the pyrene units. TP-COF is electrically conductive, which is ascribed to the eclipsed stacking structure of the p-electronic components. It guarantees the hole transport and is capable of on–off switching of the electric current as revealed by the I–V curve measurements (Fig. 6.5).

Because phthalocyanine derivatives have abundant π-electronic cores, the synthesis of π-electronic COFs has been achieved by the introduction of metallophthalocyanine units into the skeletons. For example, the condensation of [(OH)8PcNi] with BDBA produces a 2D NiPc-COF (Fig. 6.6) [14]. NiPc-COF adopts a flake-like morphology and consists of eclipsed 2D sheets. The test reveals that NiPc-COF is hole-conducting and with a high carrier mobility of 1.3 cm^2 V^{-1} s^{-1}. Except for the p-type NiPc-COF, 2D-NiPc-BTDA-COF [15], a nickel phthalocyanine COF using the electron-deficient benzothiadiazole as the linker (Fig. 6.7), indicated different charge carrier–transport behaviors, which is an n-type semiconductor with electron mobilities as high as 0.6 cm^2 V^{-1} s^{-1}.

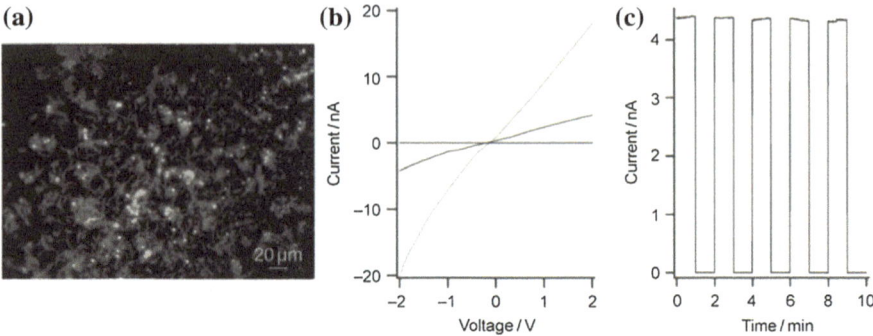

Fig. 6.5 a Fluorescence image of TP-COF; **b** (a) I–V profile of a 10-mm-wide Pt gap (*black curve* without TP-COF; *blue curve* with TP-COF; *red curve* with iodine-doped TP-COF; **c** Electric current when 2 V bias voltage is turned on or off. Reprinted with permission from Ref. [13]. Copyright 2008, Wiley-VCH

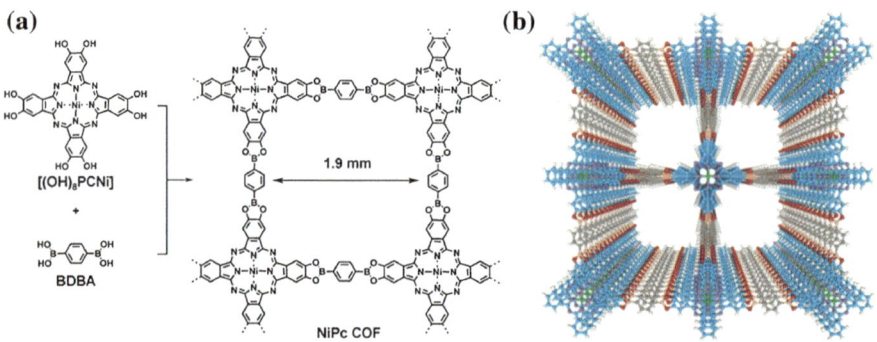

Fig. 6.6 a The synthesis of the nickel phthalocyanine covalent organic framework (NiPc-COF) by a boronate esterification reaction; **b** eclipsed stack of phthalocyanine 2D sheets and microporous channels in NiPc-COF (a 2 × 2 grid is shown). Colors used for identification: phthalocyanine unit: *sky blue*; Ni *green*, N *violet*, C *gray*, O *red*, B *orange*, H *white*. Reprinted with permission from Ref. [14]. Copyright 2011, Wiley-VCH

Porphyrin is another representative macrocycle with extended π conjugation, thus promoting the synthesis of porphyrin contained COFs. The studies indicated that the moving of charge carriers in 2D MP-COFs depend not only on the skeletons but also on the central metals in the macrocycles. The 2D porphyrin COFs containing different central metals (MP-COFs) COFs show a tunable charge carrier transport quality (Fig. 6.8) [16]. When inserting a copper central metal ion, CuP-COF leads to an electron transport along the framework. Meanwhile, the metal-free porphyrin COF is hole-conducting. By contrast, the ZnP-COF shows ambipolar charge transport properties. Flash-photolysis time-resolved microwave conductivity (FP/TRMC) methods under argon atmosphere indicate the total capability of carrier (electrons and holes) mobilities.

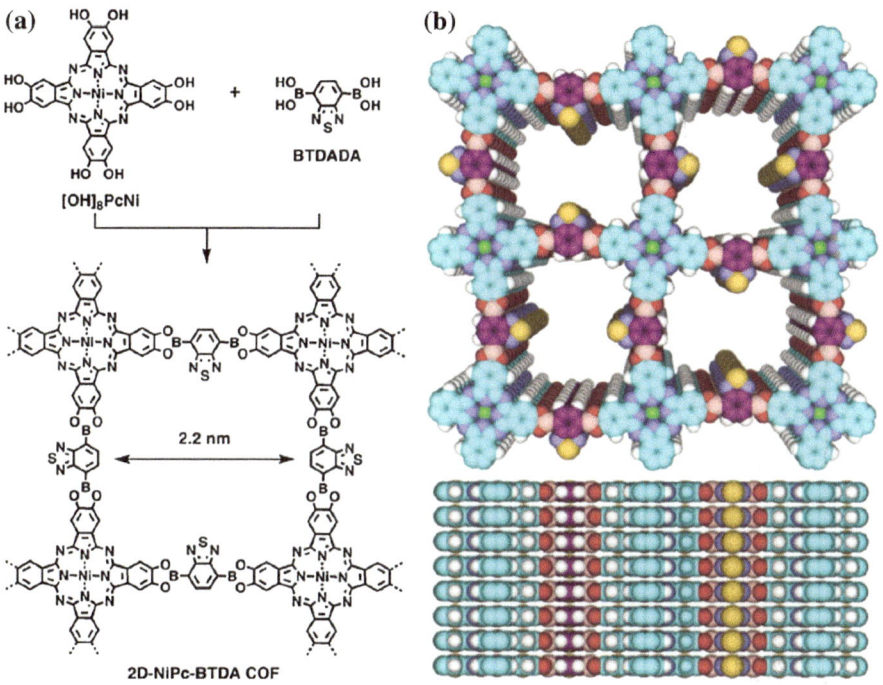

Fig. 6.7 **a** Schematic representation of the synthesis of 2D-NiPc-BTDA COF with metalloph-thalocyanine at the vertices and BTDA at the edges of the tetragonal framework; **b** *top* and *side views* of a graphical representation of a 2 × 2 tetragonal grid showing the eclipsed stacking of 2D polymer sheets (Pc *sky blue*; BTDA *violet*; Ni *green*; N *blue*; S *yellow*; O *red*; B *orange*; and H *white*). Reprinted with permission from Ref. [15]. Copyright 2011, American Chemical Society

Other two porphyrin COF materials (COF-66 and COF-366) reported by Yaghi et al. are hole-conducting and their mobilities reach as high as 8.1 and 3.9 cm^2 V^{-1} s^{-1}, respectively (Fig. 6.9) [17]. It is notable that the interlayer spacing in COF-366 (5.64 Å) is much larger than that of COF-66 (3.81 Å). However, the carrier mobility of COF-366 is twice that of COF-66, suggesting that the charge carriers can move not only through the stacking layers but also over the intralayer planes in the π-conjugated imine-linked plane.

6.3.3 Photoconduction

The development of photofunctional materials has attracted much attention thanks to their important applications in the fields of artificial photosynthesis, light energy conversion, and optoelectronics. The molecular ordering of the π electronic components is important, which could influence the performance of these devices.

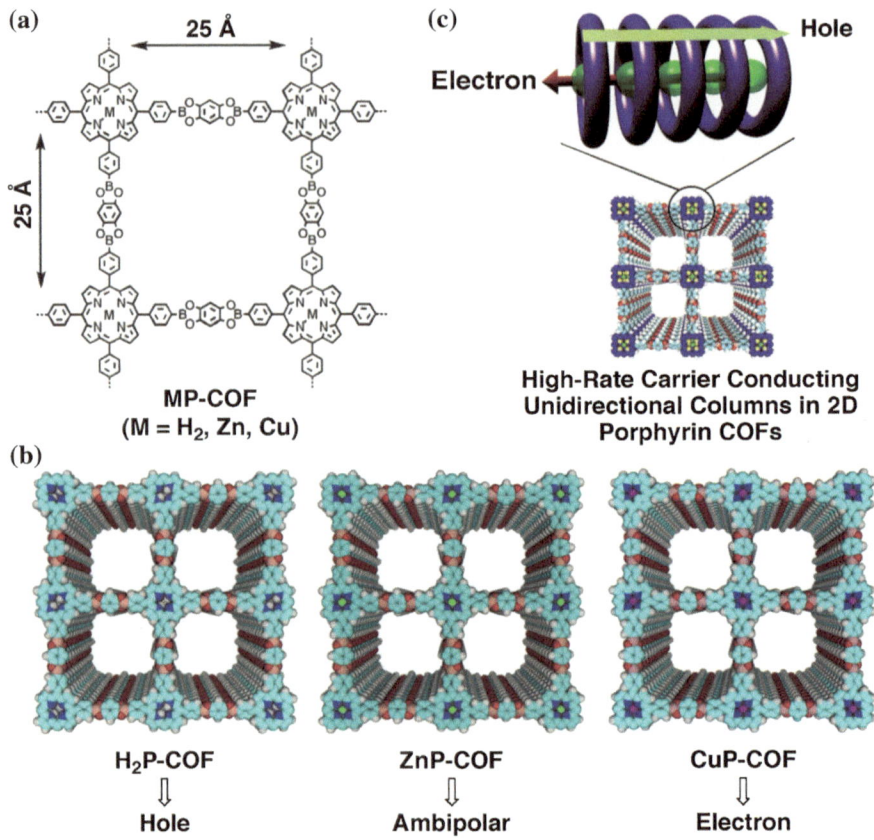

Fig. 6.8 **a** Schematic representation of MP-COFs (M=H$_2$, Zn, and Cu); **b** schematic graphs of a 2 × 2 grid of MP-COFs with achiral AA stacking 2D sheets (C *light blue*; N *deep blue*; H *white*; O *red*; B *pink*; Zn *green*; Cu *violet*); **c** Graphical representation of metal-on-metal and macrocycle-on-macrocycle channels for respective electron and hole transport in stacked porphyrin column of 2D porphyrin COFs. Reprinted with permission from Ref. [16]. Copyright 2012, Wiley-VCH

It has been documented that photoconductivity can be achieved using high-quality single crystals of certain π-conjugated arenes. First, the exciton migrates over the lattice, and then the charge will separate at the molecule–electrode interface. COFs is endowed with a high probability for triggering photoconductivity due to the eclipsed stacking structure and periodic alignment of the π columns.

Jiang reported the first example of a photoconductive COF, which is a pyrene-based COF synthesized via the self-condensation of PDBA (Fig. 6.10) [18]. Fluorescence measurements reveal that the PPy-COF is highly blue luminescent because of the formation of excimer in the stacked pyrenes. PPy-COF quickly responds to the irradiation with visible light to generate a prominent photocurrent (Fig. 6.11).

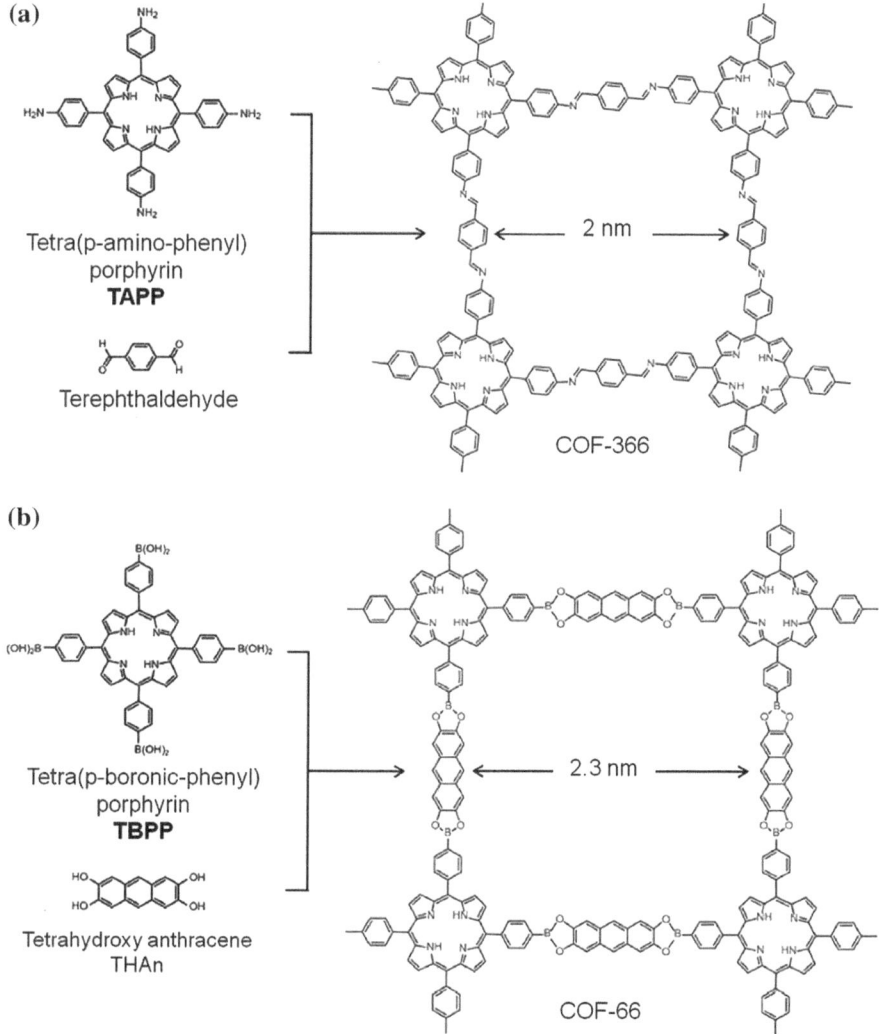

Fig. 6.9 Condensation reactions between TAPP and Terephthaldehyde, TBPP, and THAn produce extended **a** COF-366 and **b** COF-66. Reprinted with permission from Ref. [17]. Copyright 2011, American Chemical Society

The studies also reveal that metalloporphyrin COFs are panchromatically photoconductive. It is similar with the phenomenon in semiconduction, the photoconductivity of metalloporphyrin COFs is highly determined by the metal species of the porphyrin macrocycles [16]. ZnP-COF shows the highest on–off ratio with a value of 5×10^4, which is 150 and 10,000 folds higher than that of H$_2$P-COF and CuP-COF, respectively (Fig. 6.12). A balanced and ambipolar charge carrier transport is crucially responsible for generating large photocurrents.

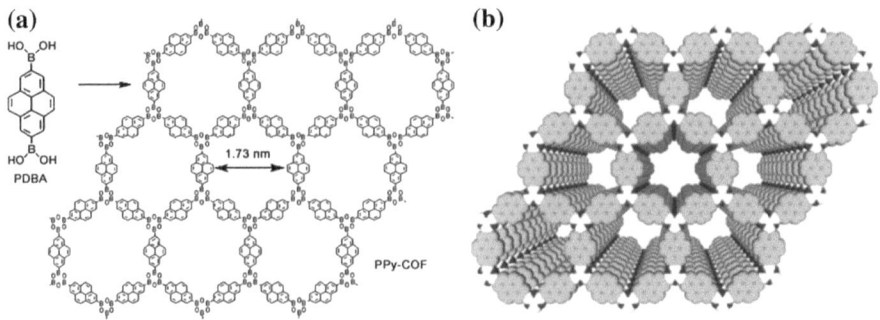

Fig. 6.10 **a** Schematic representation of the synthesis; **b** the structure of PPy-COF. The structure is based on quantum-chemical calculations and crystal lattice parameters (B *white*, O *red*, pyrene *blue*; H atoms are omitted for clarity). Reprinted with permission from Ref. [18]. Copyright 2009, Wiley-VCH

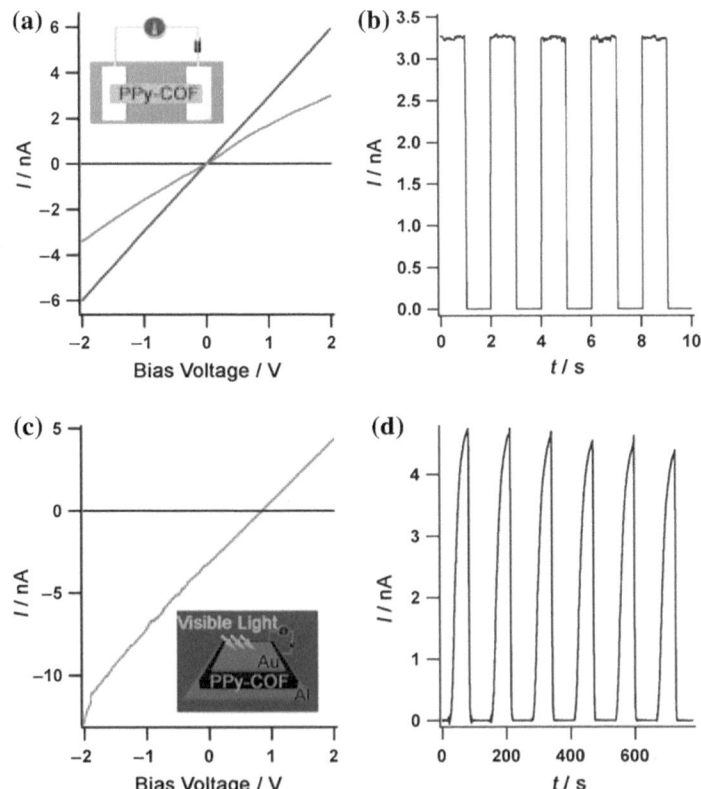

Fig. 6.11 **a** I–V profile of PPy-COF between two Pt electrodes 10 mm apart (*black curve* without PPy-COF; *blue curve* with PPy-COF; *red curve* with iodine-doped PPy-COF); **b** Electric current when the 2 V bias voltage was turned on or off; **c** I–V profile of PPy-COF between sandwich-type Al/Au electrodes (*black curve* without light irradiation; *red curve* upon light irradiation); **d** Photocurrent when the light was turned on or off

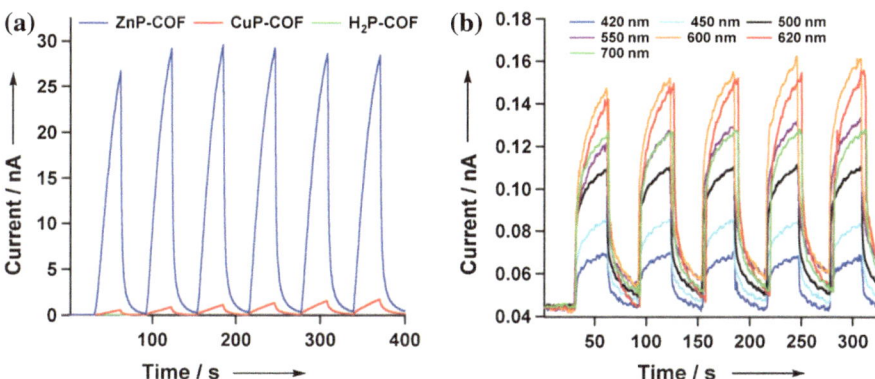

Fig. 6.12 **a** Photocurrents for 2D porphyrin COFs upon repeated switching of the light on and off; **b** Wavelength-dependent on–off switching of photocurrent of ZnP-COF at a bias voltage of 1.0 V. Reprinted with permission from Ref. [16]. Copyright 2012, Wiley-VCH

6.4 Solar Cells

A solar cell (also called a photovoltaic cell) is an electrical device that converts the energy of light directly into electricity by the photovoltaic effect. It will lead to an energy revolution. Nowadays, inorganic solar cells, organic and inorganic solar cells, and hybrid organic–inorganic solar cells have attracted worldwide attentions. Classical hybrid solar cells are comprised of nanostructures (e.g., arrays of nanorods or nanotubes) of zinc oxide, titanium oxide, or cadmium selenide which serve as acceptors for interpenetrating polymerdonors [19–21]. The composition, structure, and metrics of the synthesized POFs can be systematically varied, achieving expected functions and properties. The extraordinary properties including functional groups control of the surface, high surface areas, and stability in combination with semiconducting frameworks make them potential candidates for optoelectronics. Herein, POFs are designed and decorated with functional groups, providing a possibility to obtain an excellent photovoltaic property.

In 2013, Bertrand group reported the new family of thiophene-based COF by connection of thiophene with HHTP [22]. It is easy to form the bent skeleton using the nonlinear thiophene monomer (H$_4$TDB). Therefore, they designed and synthesized a linear thiophene (H$_4$BTDB) for construction of COFs. Subsequently, Bein et al. prepared a COF-containing stacked thieno[2,3-b]thiophene-based building blocks, with high surface area and a 3-nm open pore system (Fig. 6.13) [23]. The resulting materials could be served as electron donors. Besides, the pore size of TT-COF is very suitable to take up the well-known fullerene electron acceptor [6,6]-phenyl-C$_{61}$-butyric acid methyl ester (PCBM). The formation of novel structurally ordered donor–acceptor network exhibited significant photocurrent upon irradiation. The complex achieves a novel donor–acceptor system, and shows an efficient charge transfer.

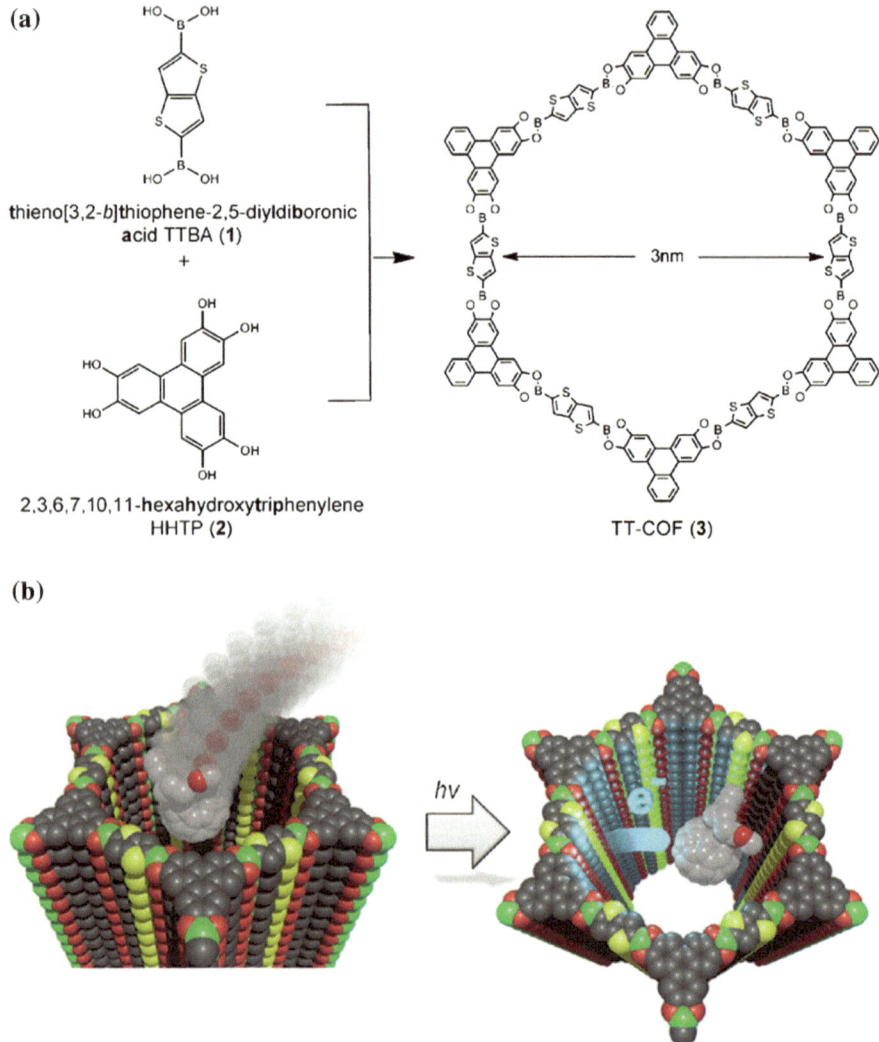

Fig. 6.13 a Reaction scheme for the co-condensation of TT-COF; **b** schematic representation of the host–guest complex of TT-COF and a PCBM molecule to scale (C *gray*, O *red*, B *green*, and S *yellow*. For clarity only one PCBM molecule is shown, whereas in the experiments described, the COF channels are loaded with the PCBM phase. Reprinted with permission from Ref. [23]. Copyright 2013, Wiley-VCH

In the same year, Jiang's group reported a chemically stable, electronically conjugated 2D COF with open nanochannels centered at 1.6 nm (Fig. 6.14a) [24]. The CS-COF allows inherently periodic ordering of fullerene in their structure. Under air atmosphere conditions, the thin film cells composed of CS-COF and C_{60} exhibited a photo conversion efficiency of 0.9 % (Fig. 6.14b). The cells possess large

(a)

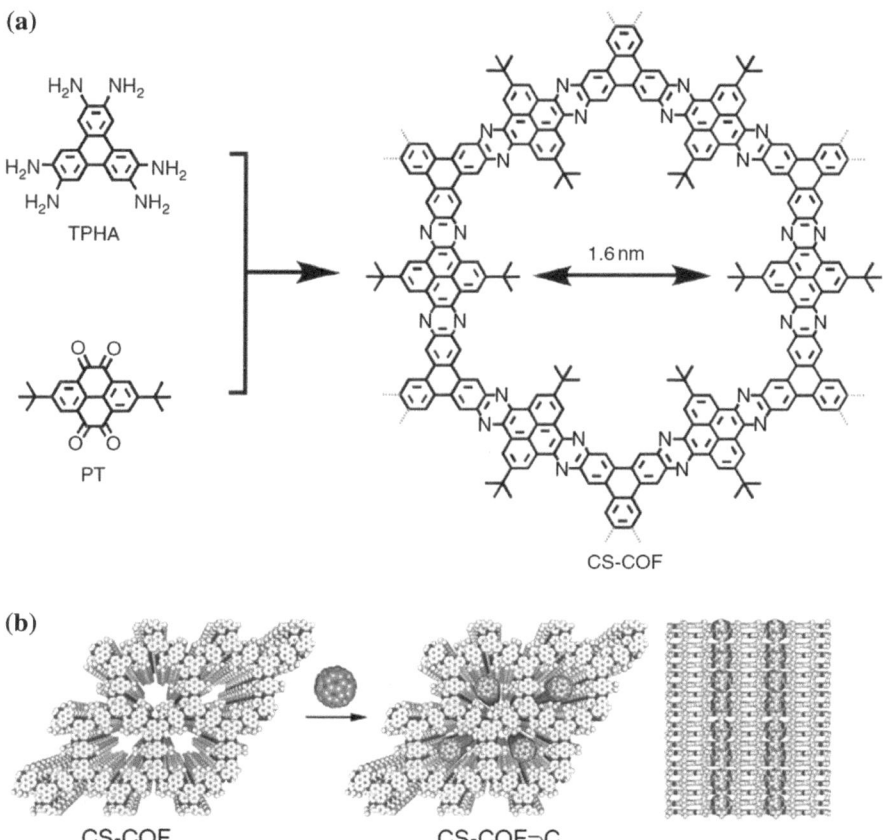

(b)

Fig. 6.14 a Schematic representation of CS-COF. b Schematic representation of CS-COF/C60 in the open one-dimensional channels. Reprinted by permission from Macmillan Publishers Ltd.: Ref. [24], copyright 2013

open-circuit voltage of 0.98 V, which is remarkable and originates from the low HOMO level of CS-COF.

6.5 Proton Conduction

An ordered one-dimensional nanochannel could conduct protons. However, the instability of covalent organic frameworks (COFs) in acid and alkali greatly limit their practical implementations. In 2014, Jiang's group designed a new strategy to gain the acid-stabilization structure (Fig. 6.15) [25]. As designed, the targeted COFs overcome their weak point and could be applied to proton conduction for the first time. The proton conductivity of Tp-Azo COF is a little lower than those of highly proton-conducting MOFs.

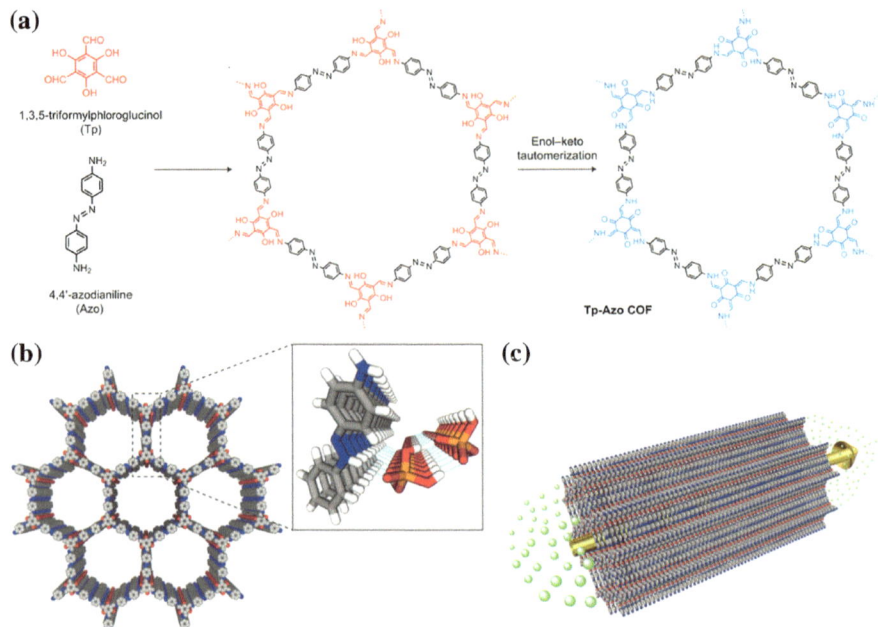

Fig. 6.15 **a** The structure of Tp-Azo COF. **b** Tp-Azo COF showing the hexagonal one-dimensional channels. **c** Graphical representation of proton conduction along the 1D channels. Reprinted by permission from Macmillan Publishers Ltd.: Ref. [25], copyright 2014

6.6 Antibacterial Performance

In 2013, our group first reported quaternary pyridinium-type PAFs, denoted as PAF-50 (Fig. 6.16) [26]. It was prepared via condensation of 4-pyridinylboronic acid and cyanuric chloride. The targeted PAFs possessed the cationic centers of quaternary pyridinium ions. The unprecedented antimicrobials PAF-1 is known to be nontoxic to mammalian cells and antibacterial. Because the pores of PAF-50 contain Cl^- as counter-ions and their sizes (5 Å) are sufficiently large to accommodate Ag^+ ions with kinetic diameter of 2.52 Å, we were encouraged to produce AgCl-loaded PAF-50. After loading AgCl inside their micropores (Fig. 6.16d), AgCl-PAF-50 integrates the excellent antimicrobial properties of quaternary ammonium and AgCl (Fig. 6.17). Besides, the aromatic framework exhibits the excellent compatibility with various organic solvents and conventional polymers. The complex of AgCl-PAF-50 and polymer leads to great operation ease and flexibility for large-scale antibacterial coating on the surfaces of diverse medical devices simply via solution coating or spray.

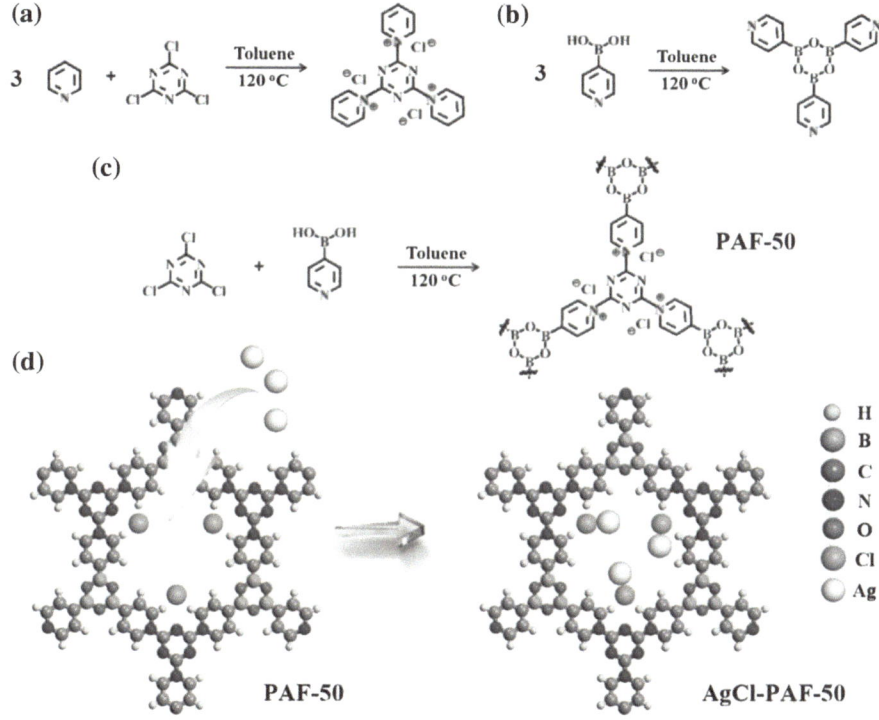

Fig. 6.16 Condensation reactions of discrete molecules and extended PAFs (**a**, **b**). Scheme of synthesis of PAF-50 (**c**) and AgCl-PAF-50 (**d**)

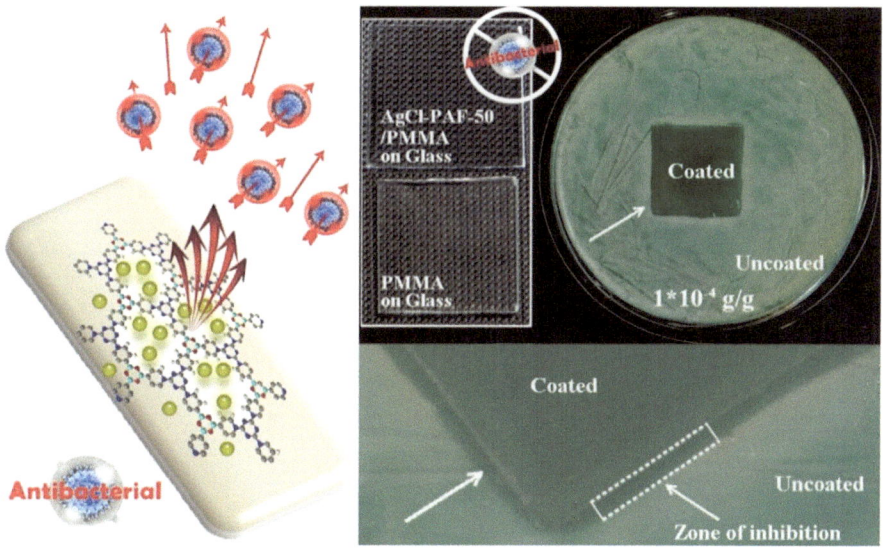

Fig. 6.17 Photo of the growth of S. aureus in Petri disks overnight, where a glass slide coated with AgCl-PAF-50/PMMA composite with the AgCl-PAF-50 to PMMA mass ratio of 1×10^{-4}

6.7 POFs for Iodine Capture

Due to its ultrahigh energy density and low carbon emissions, nuclear power is a promising and widely utilized energy source. One urgent problem is the residual nuclear waste containing radioactive materials. Thus appropriate technologies should be developed for treating the wastes [27]. The ^{129}I and ^{131}I are two of the main components of waste streams. It is notable that the half-life of ^{129}I reaches 15.7 million years. They can enter into our body through food and water, especially ^{129}I would affect human metabolic processes throughout our whole lives. The methods, such as wet scrubbing and capturing iodine with porous materials, have been developed. To stratify the requirements of this application, the materials should be stable, effective, and with high-iodine adsorption capacity. POFs featuring high surface areas and physicochemical stability could be considered as promising candidates.

Very recently, Ben and Qiu et al. demonstrated using PAF-1 [28] and JUC-Z2 [29] for the capture of iodine (Fig. 6.18) [30]. They showed ultrahigh capacity for iodine adsorption, at 298 K and 40 Pa. The iodine vapor uptake of PAF-1 and JUC-Z2 was 1.86 and 1.44 g g^{-1}, respectively. In addition, PAF-1 and JUC-Z2 could adsorb iodine over water with the selectivity of 5.1 and 6.5, respectively. Raman spectroscopy was carried out to confirm the structure of the iodine, revealing in fashion of I_5^- in POFs. Furthermore, different solvents, including n-hexane, chloroform, and methanol, were selected to carry out the iodine binding measurements.

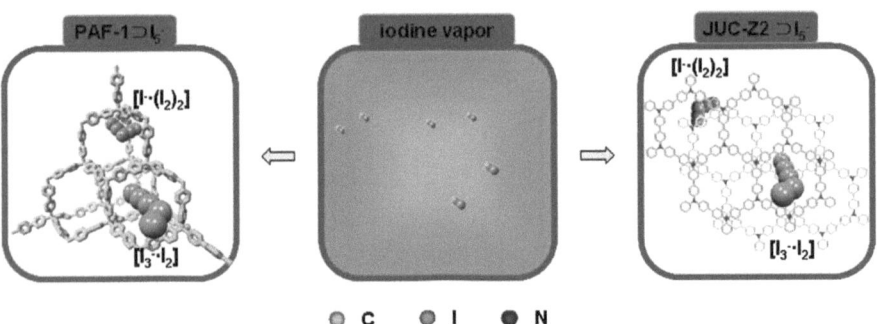

Fig. 6.18 Method for introducing iodine by diffusion in iodine vapor. Reprinted with permission from Ref. [30]. Copyright 2014, Royal Society of Chemistry

6.8 POFs for Li–S Battery

Lithium-ion batteries, with the highest energy density among all known rechargeable battery systems, are considered to be the best candidates for transportation applications. However, both the energy and power density of the lithium-ion batteries should be enhanced for extending the driving range. Thereby, many pioneering and effective strategies have been proposed on exploring high energy electrode materials in the past decades [31–33]. Because sulfur can react with lithium in the formation of Li_2S to generate the highest theoretical capacity of 1,675 mA h g^{-1}, and with an average voltage of around 2.15 V, sulfur is a very promising cathode candidate. It is notable that Lithium-ion battery can achieve a high theoretical energy density of 2,600 W h kg^{-1}.

For Li–S batteries, there are two major problems limiting their practical applications. One is the electronic insulating nature of the elemental sulfur and the discharge products, resulting in poor utilization of the active material. To resolve this problem, the effective method is to disperse sulfur into a highly conductive media, such as porous carbon [34] or a conducting polymer [35], or to use additive electron conductors such as acetylene black [36]. The other problem is its poor cycle stability that resulted from the highly soluble intermediate lithium polysulfides. Alternative electrolytes, including polymer electrolytes, glass–ceramic electrolytes, and ionic liquid electrolytes, have been attempted to enhance the cycling stability.

In 2013, Guo and Dai et al. demonstrated a simple method to prepare a sulfur cathode using an electroactive PAF as the host for lithium–sulfur batteries [37]. The PAF–S composite was prepared by loading sulfur via melt diffusion at 155 °C. As a result, the PAF–S composite exhibited high capacity and excellent cycling stability in the sulfone electrolyte of 1.0 M $LiPF_6$–MiPS and in the ionic liquid electrolyte of 0.5 M LiTFSI–MPPYTFSI. Notably, after 50 cycles in the system of 0.5 M LiTFSI–MPPY TFSI, the PAF–S composite still retained a capacity of 690 mA h g^{-1}, which is about 83 % of the initial reversible capacity (Fig. 6.19). The results suggest that sulfur

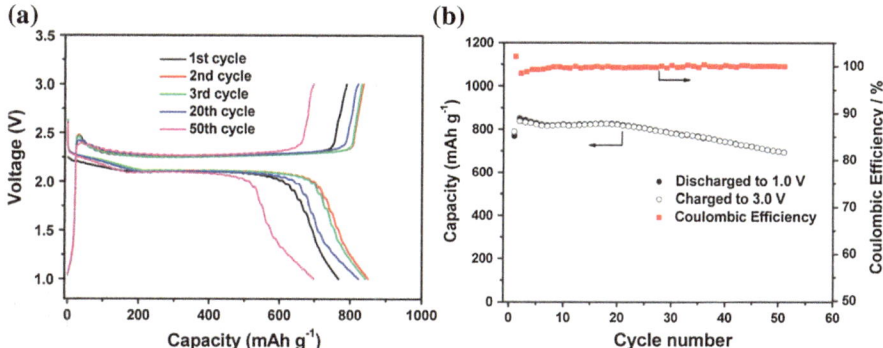

Fig. 6.19 Galvanostatic discharge–charge curves (**a**) and cycling performance (**b**) of the LiJPAF–S cell at a rate of 0.05 C in 0.5 M LiTFSI–MPPY.TFSI at 50 °C. Reprinted with permission from Ref. [37]. Copyright 2013, Royal Society of Chemistry

loaded within the pores of PAF materials leading to confinement effect, thus the poor cycle stability issue of the sulfur electrode could be partially mitigated. By proper optimization, PAF–S could be considered as a promising candidate for Li–S batteries.

6.9 POFs for Oxygen Reduction Reaction

Among the most efficient energy conversion devices, the proton-exchange membrane fuel cell (PEMFC) is considered as a promising candidate for future transportation applications [38]. The PEMFC is operated through the electrochemical hydrogen oxidation reaction (HOR) at the anode and oxygen reduction reaction (ORR) at the cathode. Thus the development of new electrocatalysts with enhanced oxygen reduction activity is crucial to satisfy the demands. Currently, the precious metals, such as platinum supported on a carbon substrate, are the most widely used electrocatalysts [39–41]. However, they suffer from the drawbacks of high cost and low stability, limiting the precious metals for large-scale commercialization. To solve this problem, a wide range of active and inexpensive nonprecious metal catalysts, including transition metal (Fe, Co) nitrogen-coordinated complexes [42, 43], pyrolyzed metal/nitrogen/carbon hybrids from two (or more) individual precursors of nitrogen and metal (Fe, Co) salts [44, 45], and conductive polymers [46], have been explored because they show promising activities toward ORR.

Among the CMPs, the metalloporphyrin (Fe, Co)-based CMPs displayed high surface area and various networks [47]. It is notable that they possess unique carbon-rich skeleton with inherent metal–nitrogen coordination. Employing these CMPs as precursor, the transition metal–nitrogen doped carbons would be obtained through the high-temperature pyrolysis.

In 2013, Liu reported a highly active and support-free oxygen reduction catalyst prepared from ultrahigh-surface-area porous polyporphyrin (Fig. 6.20) [48]. The resulting PFeTTPP powder was thermally activated under flowing nitrogen from 600 to 1,000 °C. The as-synthesized PFeTTPP showed little catalytic activity. After being pyrolyzed at 600 °C, the product started to become active toward ORR owing to the conversion of the organic species to carbon. The activity increased dramatically when the temperature reached 700 °C (PFeTTPP-700) with an onset potential of 0.93 V. The NPMC demonstrated a high onset potential and selectivity toward the reduction of oxygen to water in the RRDE test. This approach shows a new route for NPMC preparation. The N-containing functional groups could be used as the ligation sites for transition metals. Therefore, different metals can be incorporated into POPs by rational design, producing different compositions and surface properties to satisfy the demands of ORR.

Very recently, Müllen and Dai et al. reported cobalt porphyrin-based conjugated mesoporous polymers for ORR, respectively [49, 50]. The materials were synthesized by Yamamoto polycondensation (Fig. 6.21). The resulting materials show excellent catalytic activity and stability in both alkaline and acidic media. Interestingly, Dai et al. attempted the introduction of metals into the preformed

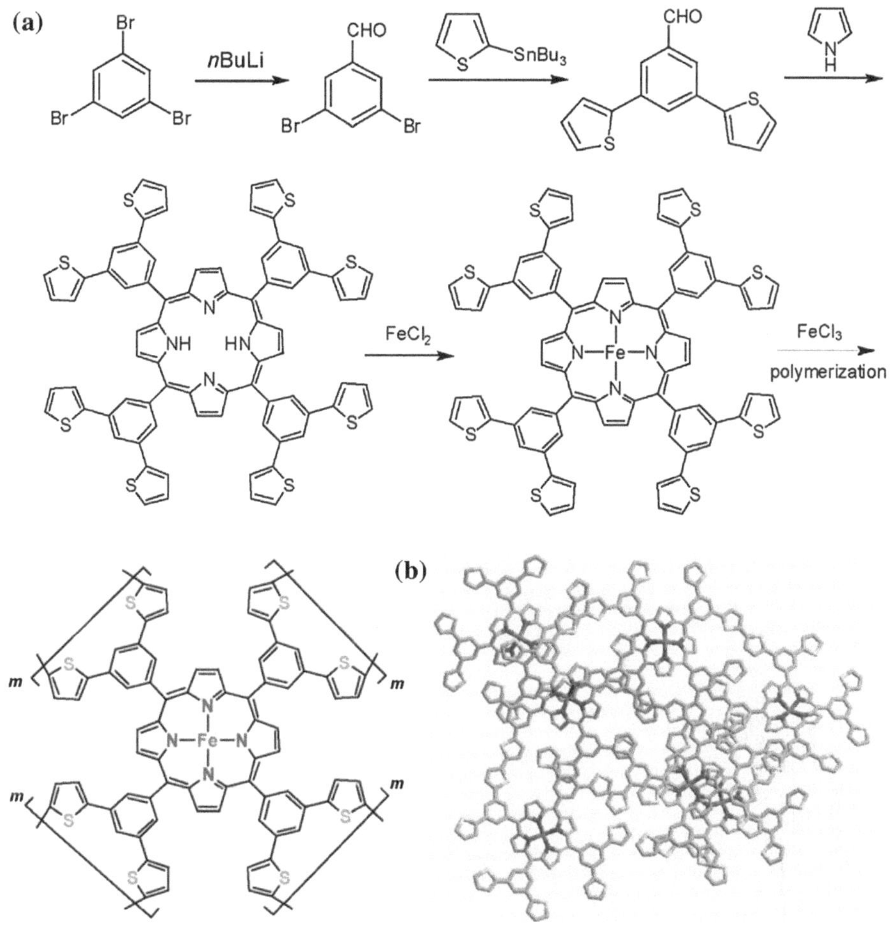

Fig. 6.20 Synthesis of PFeTTPP. **a** Molecular structure; **b** simulated 3D stacking of PFeTTPP. Fe *red*, N *blue*, C *light blue*, S *yellow*; H not shown for clarity. Reprinted with permission from Ref. [48]. Copyright 2013, Wiley-VCH

COP-P. But the resulting material showed no catalytic activity. It was resulted from the poor solubility of the synthesized COP materials in most common solvents.

Such a template-free approach toward porous carbons with controlled integration of metal nanoparticles and unique structural features opens up new avenues to nanostructured carbon materials for fuel cells, batteries, and supercapacitors.

Indeed, POFs have displayed a great deal of excellent performances thanks to their self-characteristics, including high surface area, high stability, tuned pore size, and various structures, etc. Via creative exploration, the unexpected properties of POFs could be discovered.

(a)

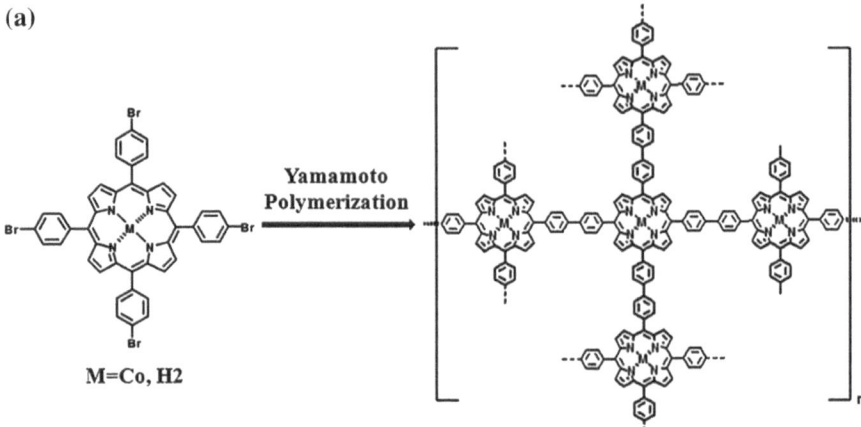

(b)

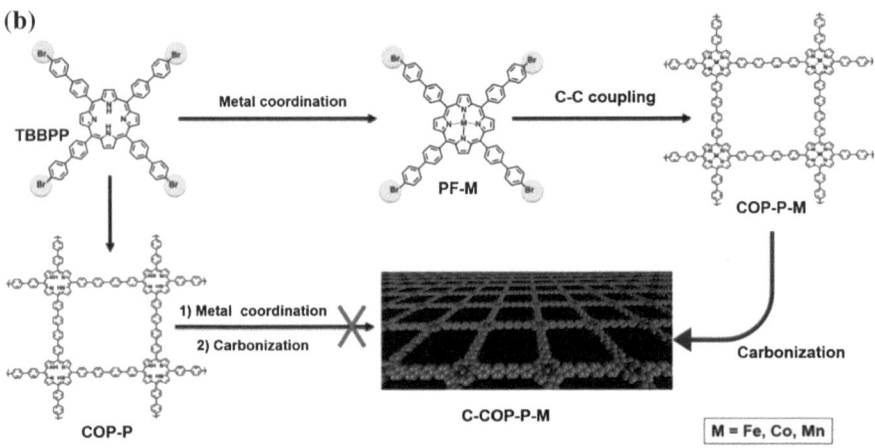

Fig. 6.21 a Schematic representation of the chemical synthesis of metalloporphyrin-based conjugated mesoporous polymer frameworks (M=Co, H$_2$); **b** the incorporation of nonprecious metals (Fe, Co, or Mn) into C-COP in this work. Owing to the poor solubility of the synthesized COP materials in most common solvents, it is too difficult to introduce metals into the preformed COP-P. Reprinted with permission from Ref. [49, 50]. Copyright 2013, Wiley-VCH

References

1. McKeown N, Budd P (2010) Exploitation of intrinsic microporosity in polymer-based materials. Macromolecules 43:5163–5176
2. Dawson R, Cooper A, Adams D (2012) Nanoporous organic polymer networks. Prog Polym Sci 37:530–563
3. Cooper A (2009) Conjugated microporous polymers. Adv Mater 21:1291–1295
4. Thomas A, Kuhn P, Weber J et al (2009) Porous polymers: enabling solutions for energy applications. Macromol Rapid Commun 30:221–236

5. Xu Y, Jin S, Xu H et al (2013) Conjugated microporous polymers: design, synthesis and application. Chem Soc Rev 42:8012–8031
6. Chen L, Yang Y, Jiang D (2010) Light-harvesting conjugated microporous polymers: rapid and highly efficient flow of light energy with a porous polyphenylene framework as antenna. J Am Chem Soc 132:6742–6748
7. Liao K, Yambem S, Haldar A et al (2010) Designs and architectures for the next generation of organic solar cells. Energies 3:1212–1250
8. Li G, Zhu R, Yang Y (2012) Polymer solar cells. Nat Photonics 6:153–161
9. Hoppe H, Sariciftci N (2004) Organic solar cells: an overview. J Mater Res 19:1924–1945
10. Lukose B, Kuc A, Frenzel J et al (2010) On the reticular construction concept of covalent organic frameworks. Beilstein J Nanotech 1:60–70
11. Patwardhan S, Kocherzhenko A, Grozema F et al (2011) Delocalization and mobility of charge carriers in covalent organic frameworks. J Phys Chem C 115:11768–11772
12. Feng X, Ding X, Jiang D (2012) Covalent organic frameworks. Chem Soc Rev 41:6010–6022
13. Wan S, Guo J, Kim J et al (2008) A belt-shaped, blue luminescent, and semiconducting covalent organic framework. Angew Chem Int Ed 47:8826–8830
14. Ding X, Guo J, Feng X et al (2011) Synthesis of metallophthalocyanine covalent organic frameworks that exhibit high carrier mobility and photoconductivity. Angew Chem Int Ed 50:1289–1293
15. Ding X, Chen L, Honsho Y et al (2011) An n-channel two-dimensional covalent organic framework. J Am Chem Soc 133:14510–14513
16. Feng X, Liu L, Honsho Y et al (2012) High-rate charge-carrier transport in porphyrin covalent organic frameworks: switching from hole to electron to ambipolar conduction. Angew Chem Int Ed 51:2618–2622
17. Wan S, Gandara F, Asano A et al (2011) Covalent organic frameworks with high charge carrier mobility. Chem Mater 23:4094–4097
18. Wan S, Guo J, Kim J et al (2009) A photoconductive covalent organic framework: self-condensed arene cubes composed of eclipsed 2D polypyrene sheets for photocurrent generation. Angew Chem Int Ed 48:5439–5442
19. Weickert J, Dunbar R, Hesse H et al (2011) Nanostructured organic and hybrid solar cells. Adv Mater 23:1810–1828
20. McGehee M (2009) Nanostructured organic-inorganic hybrid solar cells. MRS Bull 34:95–100
21. Dittrich T, Belaidi A, Ennaoui A (2011) Concepts of inorganic solid-state nanostructured solar cells. Sol Energ Mat Sol C 95:1527–1536
22. Bertrand G, Michaelis V, Ong T et al (2013) Thiophene-based covalent organic frameworks. P Natl Acad Sci USA 110:4923–4928
23. Dogru M, Kunz T, Medina D et al (2013) A photoconductive thienothiophene-based covalent organic framework showing charge transfer towards included fullerene. Angew Chem Int Ed 52:2920–2924
24. Guo J, Xu Y, Jin S et al (2013) Conjugated organic framework with three-dimensionally ordered stable structure and delocalized pi clouds. Nat Commun 4:2736–2743
25. Xu H, Jiang D (2014) Covalent organic frameworks crossing the channel. Nat Chem 6:564–566
26. Yuan Y, Sun F, Zhang F (2013) Targeted synthesis of porous aromatic frameworks and their composites for versatile, facile, efficacious, and durable antibacterial polymer coatings. Adv Mater 25:6619–6624
27. Soelberg N, Garn T, Greenhalgh M et al (2013) Radioactive iodine and krypton control for nuclear fuel reprocessing facilities. Sci Technol Nucl Ins 2013:702496
28. Ben T, Ren H, Ma S et al (2009) Targeted synthesis of a porous aromatic framework with high stability and exceptionally high surface area. Angew Chem Int Ed 48:9457–9460
29. Ben T, Shi K, Cui Y et al (2011) Targeted synthesis of an electroactive organic framework. J Mater Chem 21:18208–18214

30. Pei Y, Ben T, Xua X et al (2014) Ultrahigh iodine adsorption in porous organic frameworks. J Mater Chem A 2:7179–7187
31. Poizot P, Laruelle S, Grugeon S et al (2000) Nano-sized transition-metal oxides as negative-electrode materials for lithium-ion batteries. Nature 407:496–499
32. Hu Y, Adelhelm P, Smarsly B et al (2007) Synthesis of hierarchically porous carbon monoliths with highly ordered microstructure and their application in rechargeable lithium batteries with high-rate capability. Adv Funct Mater 17:1873–1878
33. Guo B, Wang X, Fulvio P et al (2011) Soft-templated mesoporous carbon-carbon nanotube composites for high performance lithium-ion batteries. Adv Mater 23:4661–4666
34. Ji X, Lee K, Nazar L (2009) A highly ordered nanostructured carbon–sulphur cathode for lithium–sulphur batteries. Nat Mater 8:500–506
35. Xiao L, Cao Y, Xiao J et al (2012) A soft approach to encapsulate sulfur: polyaniline nanotubes for lithium-sulfur batteries with long cycle life. Adv Mater 24:1176–1181
36. Lai C, Gao X, Zhang B et al (2009) Synthesis and electrochemical performance of sulfur/highly porous carbon composites. J Phys Chem C 113:4712–4716
37. Guo B, Ben T, Bi H et al (2013) Highly dispersed sulfur in a porous aromatic framework as a cathode for lithium-sulfur batteries. Chem Commun 49:4905–4907
38. Wagner F, Lakshmanan B, Mathias F et al (2010) Electrochemistry and the future of the automobile. J Phys Chem Lett 1:2204–2219
39. Bashyam R, Zelenay P (2006) A class of non-precious metal composite catalysts for fuel cells. Nature 443:63–66
40. Gong K, Du F, Xia Z et al (2009) Nitrogen-doped carbon nanotube arrays with high electro-catalytic activity for oxygen reduction. Science 323:760–764
41. Li Y, Zhou W, Wang H et al (2012) An oxygen reduction electrocatalyst based on carbon nanotube–graphene complexes. Nat Nanotechnol 7:394–400
42. Jahan M, Bao Q, Loh K et al (2012) Electrocatalytically active graphene–porphyrin MOF composite for oxygen reduction reaction. J Am Chem Soc 134:6707–6713
43. Chen Z, Higgins D, Yu A et al (2011) A review on non-precious metal electrocatalysts for PEM fuel cells. Energy Environ Sci 4:3167–3192
44. Lefevre M, Proietti E, Jaouen F et al (2009) Iron-based catalysts with improved oxygen reduction activity in polymer electrolyte fuel cells. Science 324:71–74
45. Wu G, More K, Johnston C et al (2011) High-performance electrocatalysts for oxygen reduction derived from polyaniline, iron, and cobalt. Science 332:443–447
46. Khomenko V, Barsukov V, Katashinskii A et al (2005) The catalytic activity of conducting polymers toward oxygen reduction. Electrochim Acta 50:1675–1683
47. Chen L, Yang Y, Guo Z et al (2011) Highly efficient activation of molecular oxygen with nanoporous metalloporphyrin frameworks in heterogeneous systems. Adv Mater 23:3149–3154
48. Yuan S, Shui J, Grabstanowicz L et al (2013) A highly active and support-free oxygen reduction catalyst prepared from ultrahigh-surface-area porous polyporphyrin. Angew Chem Int Ed 52:8349–8353
49. Wu Z, Chen L, Liu J et al (2014) High-performance electrocatalysts for oxygen reduction derived from cobalt porphyrin-based conjugated mesoporous polymers. Adv Mater 26:1450–1455
50. Xiang Z, Xue Y, Cao D et al (2014) Highly efficient electrocatalysts for oxygen reduction based on 2D covalent organic polymers complexed with non-precious metals. Angew Chem Int Ed 53:2433–2437